American Building

American Building

The Chicago History of American Civilization
Daniel J. Boorstin, Editor

Carl W. Condit

Materials and Techniques
from the First Colonial Settlements to the Present

Second Edition

The University of Chicago Press
Chicago and London

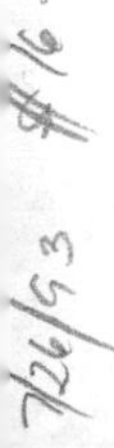

The University of Chicago Press, Chicago 60637
The University of Chicago Press, Ltd., London

 Published 1968
Second edition 1982
Printed in the United States of America

89 88 3 4 5

Library of Congress Cataloging in Publication Data

Condit, Carl W.
American building.

(The Chicago history of American civilization)
Bibliography: p. 295.
Includes index.
1. Building—United States—History. I. Title.
II. Series.
TH23.C58 1982 690′.0973 82-13602
ISBN 0-226-11448-1
ISBN 0-226-11450-3 (pbk.)

Editor's Preface

Much of our study of the buildings of our past has been (in the literal sense of the word) superficial. We have found it much easier to look at the surface, describing the shape and ornament and color and texture of the outside of American buildings, than to understand what they really were made of and what held them together. In our own age, when we see buildings go up, we cannot help knowing something of what lies underneath the surface. But when we look at the buildings from the past we usually see nothing but the finished product. Only when the outer coverings have come off, sometimes quite by accident, do we begin to see the structure within. In this pioneer book, Mr. Condit helps us imagine what it must have been like to watch the construction of a colonial half-timbered house or to witness the erection of the national Capitol.

Mr. Condit teaches us primarily the anatomy and the physiology —rather than the cosmetics—of American buildings. Although everything he tells us reinforces the intimate connection between how people make a building and how they make it beautiful, his focus is on what holds the building together. Since he is here primarily interested in materials and techniques of construction, he gives us an emphasis not commonly found in histories of architecture, by means of a view of American bridges, dams, and other structures in which the building elements are revealed in their nakedness. He shows us the connections between the forms of our bridges and locks and railroad train sheds and the buildings in which we live and work. In this and others ways he opens our eyes and helps us learn more from the construction that we hear and see today everywhere in America. This book, then, should not only help us understand better what we have already seen but should help us learn to observe more sharply.

We have often had explained to us how peculiarly American problems and opportunities and passions brought into being the more spectacular American buildings, such as the skyscraper. And Louis H. Sullivan and others have shown us how the needs of these buildings in turn led to revolutions in the uses of materials and even in the very function of architecture. Mr. Condit now explores the context of many less spectacular innovations in American construction. He shows us how the needs and limits of American life induced

Americans to import, to adapt, and to invent. He takes us from the primitive cellars, dugout shelters, and "English wigwams" of the earliest settlers, through the log cabin (a Swedish and not a frontier invention), down to the most recent uses of aluminum, steel, and concrete, and back in full circle to the newest uses of one of the oldest of building materials, wood, which has been given surprising new forms.

Among the heroes of his story are wood and steel and concrete. Mr. Condit admirably serves the purposes of the Chicago History of American Civilization by drawing all these elements—and the history of the arts and materials of construction—into the main stream of American history. The series contains two kinds of books: a *chronological* group, which provides a coherent narrative of American history from its beginning to the present day; and a *topical* group (including the present volume) which deals with the history of varied and significant aspects of American life, aiming to make each aspect of our culture a window to all our history.

DANIEL J. BOORSTIN

Contents

Part One

Colonial Building Development

Part Two

The Agricultural Republic: Revolution to Civil War

Part Three

The Rise of the Industrial Republic, 1865–1900

Part Four

Industrial and Urban Expansion in the Twentieth Century

List of Illustrations

List of Illustrations

List of Illustrations

List of Illustrations

Part One

Colonial Building Development

Chapter One | Timber Framing

The first English settlers to land on the American shore were for the most part rudely educated and simple people who had played little part in the high culture of Jacobean England. Small farmers, artisans, and *petite bourgeoisie* from the towns of East Anglia, the Puritans who established the colony at Massachusetts Bay were patient men capable with their hands but not likely to be technical innovators. The only building art they could have known at firsthand was the medieval tradition embodied in the timber houses and barns of the small town and its surrounding fields. The colonies farther south—Maryland, Virginia, the Carolinas—were populated mainly by the same classes of yeomen and artisans, together with a number of indentured servants and a few transported felons. Lord Baltimore and his family in Maryland, and the few gentlemen of the London Company in Virginia, brought with them a higher level of taste and knowledge in architecture through acquaintance with the manors of Tudor and Jacobean estates.

The New Englanders had to struggle to maintain their precarious farms on the granite-bound soil of the region, but with tobacco as the chief commodity and the rich soil of the Piedmont at their disposal, the pioneers of the southern colonies quickly established a flourishing plantation economy. These differences in origin and economic level revealed themselves in marked regional differences in the character of building. When we recall that the other settlements of New York, Pennsylvania, the far South, and the Southwest sprang from Dutch, German, Swedish, French, and Spanish origins, we can readily understand the diversity of building techniques and forms that appeared in the various parts of colonial America.

But there were other determinants beyond those of origin, native experience, and economy. The most obvious and most formidable was the weather. The Puritans who disembarked at Plymouth Rock faced a double terror: an unknowr wilderness populated by hostile Indians and the cruelty of the New England winter. At Plymouth, England, they knew an extreme temperature range, from the average January low to the average July high, of 63 degrees. The comparable range at Plymouth, Massachusetts, is 110 degrees, with the extreme low at −12. The climate of the Tidewater and Piedmont areas is much pleasanter, but the long, hot, windless summers, growing in length and humidity as one approaches the Gulf, offer near-jungle weather for which the North European was wholly unprepared. And with humidity came the mosquito and the curse of malaria. In much of the Southwest the Spanish colonist lived in a desert, a furnace for half the year, its milder seasons punctuated by bitter storms that sweep down from the mountains and higher plateaus. To preserve a decent human habitation in the face of these grotesque exaggerations of a climate he once knew taxed the ingenuity of the colonist wherever he went

A tightly knit industrial society is no longer affected by the regional diversity of resources that played the major role in shaping the building arts of the colonial and early republican periods. The non-existence or high cost of transportation and the expenses and difficulties of quarrying meant that the builder had to use locally available materials which could be readily worked by hand-and-tool techniques. Wood was immediately available in large quantities everywhere except the Southwest, with the result that timber construction was dominant throughout most of the colonial area. Local supplies of suitable clays were extensive enough to warrant the use of brick or adobe masonry for more durable structures, but only after the economy had advanced to the point of supporting their greater cost. Dressed stone masonry laid up in mortar was rare in the colonial period, the usual form being loose rubble, which could be laid in irregular courses if cobbles or fragments of stratified rock lay close to the ground surface. Bridges were constructed almost entirely of timber piles and beams to the end of the eighteenth century. Iron, steel, and concrete as structural materials lay far in the future.

The history of American building from the beginnings of the first settlements recapitulated the whole history of the art from its neolithic origins. With immediate shelter a matter of desperate necessity, with a minimum of medieval hand tools and no trained skills, the first English colonists had to manage with the most primitive of temporary shelters. Some of these were adopted directly from Indian forms; others were cruder than anything the Indians would have tolerated. The most primitive were simply cellars or dugout shelters, depressions in the earth roofed with turf and logs. The only part with a structural character was the timber lining that shored up the earth around the sides of the excavation. These dugouts survived until the end of the seventeenth century but were increasingly confined to the poverty-stricken margins of settled areas.

Other forms of temporary shelter were less primitive and at least had the merit of keeping the occupant above ground. One of the most common throughout the English colonies was the domed wigwam of thatch or bark supported by bent branches tied at their upper ends. The settlers clearly learned the technique from the Indian. A variation on this form, the so-called English wigwam, was common at Salem, Massachusetts, where several were built during the early years of the village, from 1625 to 1630. This type was constructed of thin trunks bent into a vaultlike form, tied longitudinally, and covered with reeds. Still another variation on the basic technique was the Indian tent, in which the branches were arranged in a conical form and covered with bark or skins. A more substantial shelter was the log tent made from trunks disposed like the rafters of a gable roof and covered with moss or clay. A more durable kind of covering was the wattle and daub, in which thin flexible sticks were interwoven into the stouter branches of the structure and the interstices then filled with mud or clay. None of these forms is a truly framed construction, since the branches were fixed only by burying one end in the ground, and connections were made by tying or interweaving. The result was a flimsy and non-rigid structure which stood at the mercy of wind and snow.

Fortunately, these primitive conditions lasted only a few years. By 1630 the English settlements included a good proportion of artisans either trained in the techniques of medieval carpentry or able to revive them through necessity and the recollections of their homeland experience. Tools began to be imported in quantity from England, and the establishment of the power-driven sawmill at Jamestown in 1625 greatly increased the output of squared framing timbers. The carpenter's tools, like the first timber building forms, were entirely of medieval origin. Implements for cutting and shaping included the felling ax, the frow ax for splitting logs, the single-handed and the two-handed pit saw, the adz, plane, chisel, and gouge. Punching and driving tools were the hammer, drill, awl, and punch or drift pin. For accurate squaring and fitting there were the square, rule, compass, plumb, and prick or marking iron. The rough dried skins of various fish served in lieu of sandpaper; the use of animal tissue to make primitive adhesives, however, came later. Most of these tools can be traced to preclassical antiquity, and all are in use today.

With adequate tools, the memory of building forms made it possible for the colonist to try his hand at the construction of crude frames of hewn planks or squared timbers covered by sawn boards laid flush or as lapped clapboards. The record indicates that he was doing so as early as 1611. Direct experience with the more advanced forms of medieval framing soon led to the development of more elaborate structural techniques. The simplest medieval form was the cruck frame, which originated in the primitive method of supporting the ridgepole of a gabled roof on the forked ends of two posts driven into the ground. By the fourteenth century the crotchet or forked post had been replaced by two timbers hewn to a curve such that they formed a rounded inverted V with crossed ends when driven into the ground. Although there is little evidence of cruck framing in the colonies, it is clear that crotchets were employed in the early years of the Virginia settlements from the description in Captain John Smith's *Advertisements for the Unexperienced Planters of New England* (1631).

A number of structural developments lay behind the large timber-framed house of medieval England, which had reached its maturity by the fifteenth century (Fig. 1). It was more elaborate than anything the colonists were to build, but it served as the prototype for all the forms in the English colonies. Several elements of the medieval frame either did not appear or were used in greatly simplified form in the American work. Chief of these were the numerous bracing members and the embryonic trusses of the gable frames (usually a triangle of two rafters and a tie beam with a post, known as a king-post, placed at the altitude). The short horizontal cross-pieces between the king-post and the rafters in the English frame were called wind beams. They were seldom used in the American counterparts, although there are a few references to wind beams in the colonial records. The medieval builder understood the necessity for rigid bracing against wind loads. Indeed, it seems likely that the flying buttresses of the Gothic cathedral functioned partly for this purpose.

The colonial framed house, also called a *fair house* or *English house,* appeared early in the seventeenth century. It is mentioned by John Smith and was among the first group of permanent dwellings built at Salem between 1628 and 1630. The structural frame was covered with clapboard siding, which is also of medieval English origin. The open spaces between framing members in the wall were filled with clay or mud reinforced with straw or sticks. Later, in the more expensive homes, this infilling (called nogging) was brick laid up in mortar. By the eighteenth century the builders adopted the practice of using wood sheathing as the nailing base for the clapboards, with the result that the clay or brick nogging could be abandoned.

The frame itself was originally an unbraced and simplified version of its late medieval antecedent (Fig. 2). Although this system of framing was most common in New England, where timber construction enjoyed the most vigorous life, it was widely used in all the English colonies. The primary members were the sills, corner posts, and a variety of girts (girders). The posts were fixed to the sills, which in turn rested on a stone foundation of loose rubble. Closely spaced joists spanning between the girts carried the floor boards. The roof sheathing was nailed directly to the rafters, which sloped upward from the plates on top of the posts to the ridge line. By the end of the seventeenth century the developed frame included diagonal braces, usually set in the corners between posts and sills, a ridge beam running the length of the roof and carrying the upper ends of the rafters, and purlins extending along the long dimension of the roof and serving as a nailing base for the roof sheathing.

The abandonment of clay and brick nogging between the posts required the addition of studs to the wall frame. These are light boards set vertically and framed into the sills and girts to provide support for the wall sheathing. The outer wall covering was nearly always clapboard siding in the English colonies. After the Dutch introduced the shingle covering for walls about 1650, on Long Island, the shingle house migrated to New England. The shingle roof was used in all the colonies before the mid-century. The more durable but more expensive slate roof was introduced in Boston in 1654, and tiles were used before the end of the century in the more elegant and costly houses.

The structural members were tightly joined by means of a variety of devices for framing one member into another. The simplest joint was a notch cut into the side of the post to receive the end of the girt, or into the girt to take the end of the joist. The notch was carefully cut to the size of the member it was designed to receive. The dovetailed joint was used to make a tight connection between larger, more heavily loaded members, such as the girts. In this type, a keystone notch was cut in one girt to match the dovetail end of the other, the diagonal cuts making it impossible for the load to pull one member away from the other. The most secure joint, however, was the mortise-and-tenon, which was commonly used to join any members subject to tensile and shearing stresses, such as the connection between the girt and the post. In the customary practice a slot or mortise was cut into the post with chisel and mallet, and the end of the girt was reduced in section to a tongue or tenon shaped to fit snugly in the mortise. The assembly was secured by passing various kinds of tapering wooden pins known as treenails through the mortised piece and the inserted tenon.

As the framing members grew larger for churches and other public buildings, the mortise-and-tenon joint became more elaborate. By the mid-eighteenth century the carpenters were using double tenons for very heavy girders, one tenon above the other with the face between the two projecting beyond the end plane of the girder in either a vertical or an inclined plane. Both devices served to reduce the shearing stress on the tenon. To reduce both shearing and bending forces under the heaviest loads knee braces were introduced into the angle at the connection of the beam and the post. These had a variety of shapes, the larger forms having been adapted from the so-called ship's knees used to join the deck beams to the ribs in the hull of a ship.

In addition to wooden fasteners, colonial builders began early in the seventeenth century to use connecting devices of iron. The oldest and simplest of these was the hand-wrought nail, which was used in the English colonies throughout the colonial period. The great variety of nails in medieval timber construction were manufactured in quantity by smiths who sold them at a certain price per hundred nails, the cost depending on the size. This practice led to the designation of nails by price, such as a threepenny nail, or a fivepenny nail. They were imported into the colonies up to the Revolution, although local production was carried on in New England and the Philadelphia area from the mid-seventeenth century on. During the period from 1790 to 1830 the hand-wrought nail gave way to the machine-cut variety, the manufacturing process being drastically altered by the introduction of power-driven cutting machinery.

In the eighteenth century other wrought-iron connections appeared, chiefly the dog, the stirrup, and the wedge. The first was a bar with the ends bent down and hammered to points so that they could be driven into the wood. It was ordinarily used as a tie set diagonally across the corners of a right-angle joint to prevent the separation of the two members. The stirrup was a heavy U-shaped strap ordinarily used at a connection involving a high concentrated load imposed by a vertical member on a horizontal. It commonly appeared in the latter part of the century to bind a post to the bottom chord of a truss, the whole assembly being drawn tightly together by means of narrow iron wedges inserted through aligned holes in the post and the stirrup.

Roofs were developed into a variety of shapes during the seventeenth century, but the framing system remained essentially the same. The only structural difference was that between the purlin and the rafter roof. In the purlin type, which was the earlier form, the longitudinal purlins lay above the rafters. The roof sheathing ran from the ridge to the plate, at right angles to the ridge line, and was nailed to the purlins. In the rafter roof, heavy purlins were framed into the principal rafters, while a secondary group of light rafters rested on the purlins. The sheathing, running parallel to the ridge line, was nailed to the light rafters.

The chief types of roof classified according to shape were the gable, gambrel, hipped, and rainbow. The gable is the oldest, simplest, and most common; the roof consists of two planes pitched uniformly on either side of the ridge. The gambrel, now rare, is a gable roof in which the plane on any one side breaks about half-way down from the ridge and falls away at a steeper pitch. It was introduced in New England about 1650, probably to shorten the rafters necessary to support the extremely high, steeply pitched Gothic roof. The hipped form is a gable in which the ends have been cut off to form triangular surfaces sloping downward toward the front and rear walls of the house; it requires the most complex framing because of the need for two systems of rafters set in planes at right angles to each other. The hipped roof is now common in the so-called ranch and split-level houses of the modern suburb. In the rainbow roof the side planes are normally bowed out into a convex curve, although the Dutch at New Amsterdam preferred to reverse the curvature and build roofs with flaring, concave surfaces. The single-pitch or lean-to roof is a special type introduced originally to cover an extension to the main volume of the house.

The straightforward column-and-beam framing of the colonial house, whether of English, Dutch, or German origin, could be easily enlarged and altered for any kind of building—barns, storehouses, churches, forts, governmental and commercial structures. The carpenter-builder followed the pragmatic tradition of the craftsman, relying on his own experience, his ingenuity, and the accumulated knowledge of his medieval forbears. For larger buildings carrying heavier loads, his solution was to increase the size and number of structural members and to introduce bracing for stability against lateral forces. The fort built at Plymouth in 1622, for example, was probably framed with heavy oak timbers and walled with thick planking. Girts and beams with mortised joints formed a horizontal frame massive enough to support a flat roof for cannon.

As the Dutch at New Amsterdam and at New Amstel in the Delaware valley grew wealthy in the fur and tobacco trade, they began, shortly after the completion of the Manhattan fort in 1626, to build substantial houses and commercial structures. Their timber construction differed from the English system only in the flaring, steeply pitched roof and the extremely massive floor beams, almost as heavy as the girts of the English house. Heavy floor beams of this kind were sometimes so long that it was necessary to reduce the clear span by inserting brackets or knee braces at the corners between posts and beams. In the Leendert Bronck house, for example, built at West Coxsackie, New York, about 1738, the bracket was a curved member framed into notches cut into the inner face of the post and the underside of the beam. Knee braces increased the rigidity of a long-span frame as well as reduced the bending forces on the beams.

The colonists who followed William Penn to Philadelphia had the advantage of a comprehensive building guide prepared by Penn himself and included in his *Information and Direction to Such Persons as Are Inclined to America* (1684). The work contains a complete description of all framing members, such as posts, girts, sills, and the like, together with "several other small pieces, as *Wind-beams, Braces, Studs,* etc., which are made of the waste Timber." Penn also gives information on the making of iron nails and on the cost of carpentry work.

The German settlers of the Philadelphia region brought other variations on the fundamental techniques of timber framing, following the medieval antecedents of their homeland. They divided the wall frame of houses and barns into a series of bays by setting the posts close together and omitting the studs of the English practice. Their chief innovation was an extensive system of diagonal bracing in which the individual brace extended across two or more bays of the wall frame. The Moravian Meeting House in the Oley Valley, Pennsylvania, built between 1743 and 1745, provides a good example of this technique.

Philadelphia by this time was rapidly approaching the position of the richest and most cosmopolitan city in the American colonies and the second largest in the British Empire. It was here, accordingly, that the colonial building craft passed from the vernacular to what we might call the protoprofessional stage. The event that signalled this transition was the establishment in 1724 of the Carpenters' Company of the City and County of Philadelphia. Based on the

Worshipful Company of Carpenters of London, which had been founded in the early fourteenth century, the Philadelphia company was thus much like the medieval guild, and the master carpenters who constituted its membership were comparable to the master builders of the guild. The organization was somewhat analogous to the modern architects' and contractors' associations, since the master carpenter of the eighteenth century, like his medieval forebear, acted as both architect and contractor.

In 1734 the Carpenters' Company began to assemble a library of English treatises on the building arts. The systematic study of these works and the practical application of the knowledge gained from them eventually transformed the members into trained professionals who knew arithmetic, geometry, surveying, and drafting and who were responsible not only for the design of a building but also for estimating its cost and supervising its construction. As a consequence, Philadelphia soon became the center of the most advanced building arts in the English colonies. By the mid-eighteenth century associations of bricklayers, stonecutters, plasterers, and various mechanical artisans were founded in Philadelphia and other large cities like New York and Boston, but none was as wealthy and as authoritative as the Carpenters' Company.

Over the years following the establishment of the library, the company developed a body of principles and rules for designing all the structural, utilitarian, and decorative elements of a building, and for estimating the cost of the various parts based on the over-all dimensions of the structure. All designing followed the canons of eighteenth-century classicism as these were set forth in the standard English treatises of the early part of the century, of which the best known, perhaps, was Batty Langley's *Ancient Masonry, both in Theory and Practice,* published at London in 1736 (the work included timber framing as well as masonry). The rules of the Philadelphia company, with associated drawings of various structural forms and other details, were printed in the *Articles of the Carpenters Company,* published at Philadelphia in 1786. These drawings include a variety of floor-framing plans, sections showing entire building frames with diagonal braces in the angles where beams and posts are joined, variations on king-post roof trusses for spans up to 60 feet, and most remarkable of all, a trusslike wall frame, extending through the floor-to-floor height of a building, for the simultaneous support of the second floor and the attic floor over spans up to 50 feet (for details of this structural technique, see under Independence Hall, p. 16).

With sophisticated techniques such as these available to them, the builders were prepared by the mid-century to face the challenge of constructing large public buildings like churches and the assembly houses of the various political units. Containing relatively large open areas within the walls, these buildings posed peculiar problems in the framing of roofs and ceilings over the interior enclosures. This was especially true in the case of churches, where it was desirable to keep the nave free of columns for unobstructed sight lines. The simplest solution was to use long, heavy rafters spanning between the ridge beam and the side walls or wall columns, and to introduce tie beams extending transversely between each pair of rafters on either side of the ridge. If the span was so great as to produce excessive flexibility in the frame, a king-post could be inserted between the junction of the rafters and the center of the tie (Fig. 1 shows king-posts in the frame). Rigidity and strength could be further increased by adding a second set of ties that would lie between the main tie beams and the ridge beam. Within this general form a number of additions and variations were possible.

The oddest example of roof framing in colonial building is the system employed in the Old Ship Meeting House, completed in 1681 at Hingham, Massachusetts, which also represents perhaps the earliest attempt to deal with the problem of framing a roof over a wide nave. The hipped roof is supported on a set of frames spanning across the 45-foot width of the building (Fig. 3). The curious system of heavy rafters, ties, and curving struts and braces constitutes a primitive truss and hence may be regarded as the beginning of truss framing in America. The long curved strut serves to brace the rafter through the short piece that connects the strut to the rafter at its mid-point. The origin of this system may have been the medieval hammer-beam truss, like those at Westminster Abbey. The curved struts of the Old Ship Meeting House look like the inverted ribs of a ship's hull, a resemblance which gave the church its name. But the colonial builder knew only very simple trusslike frames made by rafters, ties, and intermediate braces, with one or two posts (king-post or queen-post). He did not understand the action and form of the multipanel truss, as his medieval forebear also did not.

Many of the largest churches of the Georgian period had nave ceilings in vaulted form or in a combination of flat and curved surfaces. The most impressive examples among those remaining are Christ Church, Philadelphia, built piecemeal from 1727 to 1744, from James Gibbs's design of St. Martin's-in-the-Fields in London; King's Chapel, Boston, built between 1749 and 1754 after the plans of Peter Harrison; and St. Michael's Church, Charleston, South Carolina, erected between 1752 and 1762 by Samuel Cardy. These three are among the most distinguished achievements of colonial building, with Christ Church standing at the forefront because of its early date. It is a technical masterpiece in every structural detail—brick masonry of main walls and tower, highly articulated walls, truss framing of the nave, and framing of the tower and spire. The roof over the nave is carried by purlins which are framed into the upper chords of king-post trusses the lower chords of which have curved undersurfaces. Suspended from each of these undersides is a row of hangers supporting the curving ribs to which the lath and plaster of the nave ceiling is fixed. The upper chords of the trusses are let into the masonry of the piers but are given additional support at this point by large obtuse-angled ship's knees. Since the framing of the tower and spire is extremely complex, it is unfortunate that no adequately detailed drawing has been made of this elaborate system of inclined posts, braces, and the horizontal frames of the various internal floors and platforms.

St. Michael's offered comparable difficulties in the framing of the spire above its masonry tower. The gable roof over the nave, like that of King's Chapel in Boston, rests on a series of simple trusses spanning transversely between the brick side walls. The construction of the spire represents a remarkable example of colonial ingenuity in timber framing. Above the ceiling frame of the belfry rises a timber core composed of eight massive posts set in a ring and extending upward about 60 feet, nearly to the top of the spire. The ceiling frame of the arcade above the belfry supports another set of eight inward-leaning posts which carry the rafters and the planking of the spire. The roofing flares outward immediately above the arcade, thus requiring rafters that curve out at their lower ends. The belfry floor, which must carry the great concentrated load of the bells, rests on the most massive framing system of the whole church. Cardy made a good guess in this respect, since the bells were bought in England after the tower was completed.

A special structural problem comparable to the framing of towers and steeples arose in the construction of windmills, which were introduced by the Dutch on Manhattan Island in the early seventeenth century to provide power for sawmills. The first windmill in New England was built at Watertown, Massachusetts, in 1632. Among the few colonial mills that survive, the one constructed on Nantucket Island in 1746 has been preserved with some fidelity to the original construction. The fixed portion of the mill housing is an octagonal shingle-covered tower in which the chief bearing members are eight inward-leaning posts located in the corners of the octagon. Three interior levels, connected by stairways, are supported on frames of beams and peripheral girders. The first of these platforms above the ground is stout enough to support the mill stones. The gable-roofed cap of the mill is a separate structure framed with rafters and ties and supported by a ring beam (known as a curb) at the top of the tower. The cap is thus free to rotate with the change in wind direction, the rotation being accomplished by moving an extremely long sloping "tail pole" the free end of which is fixed to the hub of a cart wheel. This pole also served to brace the cap against wind loads and to oppose the overturning force of the wind throughout the height of the tower.

The working mechanism of a windmill was large enough to constitute a kind of moving timber structure. The sails, light rectangular frames on arms heavy enough to maintain rigidity, ranged from 45 to 60 feet in diameter. The moving sails rotated the tilted driving shaft (called the windshaft), which was 16 inches in diameter in the case of the Nantucket mill. The inner end of the shaft, which was tilted to hold the sails free of the tower surface, was connected to the hub of a huge toothed wheel set in the plane normal to the axis of the shaft. This wheel engaged the teeth of a wheel in the horizontal plane that transmitted the motion of the windshaft to the vertical shaft that rotated the upper or movable stone of the mill. The horizontal wheel, known as a brake wheel, is 10 feet in diameter in the Nantucket mill. The movable parts were ordinarily made of oak because of its hardness and hence its resistance to frictional wear. The fixed structure was made partly of pine, the nearly universal building wood of the seaboard colonies.

The building that sums up the state of timber construction in the most highly developed form it was to achieve in the eighteenth century is also the most famous and historically most important work of colonial architecture. The national landmark known as Independence Hall in Philadelphia was built between 1732 and 1748 as the original State House of the Colony of Pennsylvania. The author of this celebrated work is not exactly known, but reliable evidence suggests that the drawings used for construction were prepared sometime before 1735 by Edmund Woolley. The tower and the original steeple, located on the south side of the building, were added in 1753 (the timber work of the belfry and steeple has twice been reconstructed, in 1781 and 1828). The two buildings flanking Independence Hall were added after the Revolutionary War: Congress Hall (originally the County Court House) on the west, 1787–89, and the Supreme Court Building (originally the City Hall) on the east, 1790–91. The main block of the Hall is a brick structure divided into three parts by two internal brick walls, the central area being a lobby and the two enclosures on the east and west, respectively, the Assembly Room and the Supreme Court Room.

These rooms are very nearly 40 feet square on the inside and are constructed without intermediate supports, so that the framing of the second floor and attic over the open areas called for heroic ingenuity on the part of the builders to keep the space free of structural members. The ceiling and floor of any one level are supported chiefly by a two-way system of massive girders framed into each other in such a way as to form an interlocking pattern somewhat like the fireman's basket carry (Fig. 4). The girders do not span from wall to wall but nevertheless form a rigid framework that supports the joists to which the floor planking is fixed. The two-way system, composed of girders less than half the full span in length, makes possible a great reduction in the bending force and hence the tensile stress to which a single 40-foot girder would be subject. This kind of interlocking floor frame appears to have been developed by the Italian architect Sebastiano Serlio (1475–1552) and is illustrated in Batty Langley's well-studied *Ancient Masonry*. Supplementing the girders in the work of carrying the floor frame of the Hall was a vertical trusslike frame as deep as the full height from the second floor to the attic and extending across the full width of each of the two rooms along the center line. The exact form of this truss is now conjectural, but similar frameworks in the Supreme Court Building and Congress Hall offer a guide to its probable appearance. The top chord, lying in the attic floor frame, was a continuous piece resting on posts at its ends, while the bottom chord consisted of a linear succession of timbers under the second floor. There were very likely two pairs of diagonals disposed symmetrically on either side of the center line of the truss. The spaces between the chords and the diagonals were occupied by a closely ranked array of studs. Except for the area of a central doorway, this entire framework was hidden between the two plaster surfaces constituting the second-floor partition.

The roof of Independence Hall is actually a gambrel type, but the central third is a gable with so slight a pitch as to be nearly flat. The presence of this low-pitched area and the wide span of 42 feet made it impossible to use the conventional king-post truss for the primary supports. Each of the seven trusses is a modification of the basic form in which there are three posts and double top chords, the upper members functioning as roof rafters (Fig. 5). The floor and roof framing of the Supreme Court Building and Congress Hall represents refined and more efficient variations on the techniques employed in Independence Hall. The differences arise not only from the smaller size of the two flanking buildings but also from the progress of half a century in the structural arts. These various systems of timber framing are now supplemented by auxiliary steel girders that relieve the two-hundred-year-old members of a considerable part of their load.

The framing techniques brought by the French to Louisiana differed substantially from those born of the English tradition and were at first much cruder. Louisiana did not emerge as a potentially flourishing colony until 1718, when the city of New Orleans was founded. The exclusive aim of the early settlers was to found a fur-trading empire rather than to exploit the agricultural and industrial possibilities of the region. By the mid-eighteenth century the population numbered only 5,200 inhabitants, divided between French colonists (3,200) and Negro slaves (2,000). The changes in political affiliation further retarded the growth of the colony. France ceded Louisiana to Spain following the Peace of Paris in 1763, and Spain retroceded the colony to France in 1800, forestalling Alexander Hamilton's plan to take it by force. The United States purchased the immense area from France in 1803, at last opening its agricultural and mineral wealth to systematic exploitation.

The major part of French building in the Mississippi valley was heavy framed construction at first comparatively primitive in character. The earliest structural technique and probably the antecedent of most subsequent forms was known as *poteaux-en-terre*. In this system the building wall consisted of a row of stout cedar or cypress posts, hewn to flat surfaces on opposite sides, set upright in a trench, and held in place by backfilling tightly around the lengths below grade level. The posts were spaced no more than a few inches apart, the interstices being filled with clay and grass or Spanish moss. There were no footings: the posts constituted both a foundation and a wall. Rafters or horizontal beams spanning between two parallel rows of posts carried the roof planking.

Posts set directly in the ground quickly rotted, especially at their exposed lower ends, where destructive microorganisms could easily penetrate the wood. To protect them from deterioration the French builders developed the technique known as *poteaux-sur-sole*, a method of obscure origin but possibly derived from techniques used by Huron and Iroquois Indians to construct log buildings. In this system the posts rested on a timber sill (*sole* appears to be the French cognate), in turn supported by a row of flat stone blocks set in a trench. A good example of this construction is the Court House at Cahokia, Illinois, built about 1737. A more advanced variation was the *colombage-sur-sole,* or frame set on a foundation sill. This was somewhat like the English frame: heavy squared posts were joined by pegging through mortise-and-tenon joints to a timber sill, each bay of two posts and associated sill having been assembled on the ground and raised to the vertical position in the wall. This method of "tilting up" is used at the present time for large pre-assembled units of steel or concrete frames.

The most advanced structural system used by the French in Louisiana was a kind of half-timber construction known as *briqueté-entre-poteaux.* It was introduced into New Orleans by the military engineers who had charge of civic building in the colony, and was based partly on medieval precedents, partly on techniques described in later handbooks of military engineering, especially *La Science des Ingenieurs dans la conduite des Travaux de Fortification et Architecture Civile* of Bernard de Belidor (1693–1761). In this technique a brick foundation set into a trench carried a timber sill into which the heavy squared posts were framed. The posts were spaced about three feet on centers, and heavy single or double diagonal braces were set between successive pairs of posts. A plate running horizontally over the tops of the posts carried the lower ends of the roof rafters. The triangular spaces enframed by sills, posts, and braces were filled with soft brick nogging. The timbers were originally left exposed, as in Tudor English half-timber framing, but the humid climate of Louisiana caused them to rot in a few years. To prevent this the whole surface of wood and brick was covered with lime plaster and later with brick.

The best example of half-timber construction was the original Ursuline Convent in New Orleans, designed in 1727 by Ignace François Broutin, engineer-in-chief of Louisiana, and built between 1732 and 1734 (Fig. 6). Sectional drawings made by Alexandre de Batz in 1732 show a well-developed frame of massive square-section timbers. The posts were framed into the sills at the base and into plates at the second-floor and roof levels. The two diagonal braces of each bay were framed into the horizontal members at points short of the joints of posts and horizontals, a device which simplified the connections but prevented the wall frame from acting as a truss-like unit. The method of supporting the transverse floor beams is not clear from the surviving drawings. The outer ends appear to have been set into the brick nogging of the walls, while the middle lengths were additionally supported by the two intermediate rows of posts on either side of the center corridor. The roof framing was unusually elaborate: braced king-post trusses carried the purlins, which in turn supported the roof rafters. These trusses were more advanced than any fabricated in the English colonies, chiefly because the French military engineers were well trained in Renaissance building techniques.

The structural systems we have so far described have been examples of framed construction, in which the external covering, whether wood siding or brick, is a protective curtain rather than a bearing element. To combine the two functions requires that the covering be built up as a solid bearing wall, a common and very old technique in masonry which was adapted to timber construction by medieval builders. The simplest and possibly the earliest form in the colonies was the blockhouse or the garrison house, built to function as a fortified place as well as a dwelling. In the garrison house the walls were built up of logs hewn to square or rectangular section and laid up longitudinally one above the other. At the corners the logs were joined by any one of a variety of halved, mortised, and dovetailed joints, in which matching notches cut in the ends of the timbers made a tight fit. In some cases the timbers were mortised into corner posts. Garrison houses date from the early seventeenth century, although the term blockhouse (or palisadoed house) does not appear to have been used before King Philip's War, from 1675 to 1677. The war posed a serious threat to the New England colonies, with the consequence that the carpenters in some of the towns were called upon to build garrison houses large enough to shelter all the inhabitants during an Indian attack. These large structures probably embodied a combination of bearing-wall construction and interior braced framing for greater strength and support for military equipment on an upper floor.

A derivation from the garrison suitable for the domestic scale is the so-called plank house, which appeared at Plymouth as early as 1627, spreading from there to the Connecticut valley and thence to New Hampshire. The simplicity and durability of the form made it popular in north-central New England, where it continued in use up to 1860. In the fully developed house of the eighteenth century, the exterior walls and interior partitions were solid bearing structures of stout vertical planks with their ends reduced to tenons and set in slots cut in the sills and plates, the joints secured by wooden pins or treenails. The butted planks, trimmed to fit tightly together, were usually covered on the exterior with weatherboarding of clapboard siding. The openings for doors and windows were cut directly into the planks. In the case of two-story houses a bearing plate was introduced at the second-floor level to carry both the planks of the upper story and the joists of the second floor; the similar plate for the attic floor, however, was often simply pinned by means of dowels to the inner face of the planks. The thickness of the planks ranged from 1¼ to 3 inches, but the smaller size was rare because it required a supplementary framework to provide the necessary rigidity. Planking with a minimum thickness of two inches provided a rigid and durable structure while at the same time offering considerable insulation against heat loss. Its great defect was that it was wasteful of wood, since an open framework would have performed the necessary structural role with far less material.

The best known type of bearing-wall construction in wood is the log house, which became a symbol of American frontier life. Houses with walls built up of logs laid horizontally were a medieval invention of various North European peoples and by the end of the seventeenth century were common in the Scandinavian countries, Finland, and the Baltic provinces of Russia. The form was brought to America by the Swedes who established the colony of New Sweden on Delaware Bay in 1638. The earliest description of a Swedish log house comes from the journal of a Dutch traveler, Jasper Danckaerts, and is dated 18 November 1679. He thought the log house much superior to the English framed house covered with clapboard siding, probably because the thick walls of the log house had better insulating properties if the chinks were well stopped.

The Swedish built log houses in three basic types, differentiated by the form of the timbers: the wall might be built of squared logs, split logs with the flat face set vertically, or round logs. The extent to which the logs were carefully hewn to flat surfaces and straight edges determined the tightness of the wall. The interstices between logs were stopped with a daubing of mud or clay or with a chinking of oak chips and clay. The rigidity and strength of the wall depended on the weight of the logs and the security of the joints at the corners. As a matter of fact, the corner joints provided all the rigidity the structure possessed, for it was through them that horizontal wind and snow loads had to be transmitted to the walls in which the axes of the logs lay in the direction of the pressure. Two walls thus acted as bracing or buttressing for the wall exposed on the windward side. For this reason the carpenter paid close attention to the corner construction: he used a variety of mortised, dovetailed, and notched joints for square-hewn logs; semicircular notches for the round variety; and interlocking keys or tenons set in deep vertical grooves for split logs. The door and window openings were sawed out after the wall was built up to the top of the particular opening. The opening was then framed in planks or boards fixed to the ends of the logs with wooden pegs. The roofs were usually of shingles laid on sheathing nailed to the rafters. The fireplace was rubble stone set in clay.

The Germans who began to come to Pennsylvania in the early eighteenth century brought their own variant of the log house, which was nearly always constructed of square-hewn timbers with dovetailed joints. The rapid spread of the log house, in fact, was effected mainly by Germans, who came in large numbers throughout the eighteenth and nineteenth centuries. The Scotch-Irish settlers who moved into the southern mountain region and the Ohio valley learned the technique of log-cabin construction from the Germans. By spreading it still farther, they helped to establish the myth of an indigenous American invention.

The techniques of heavy timber framing appropriate to buildings were adapted by the colonial carpenters to the construction of bridges.

There was relatively little need for permanent spans in the colonies: the movement of goods and people between settled areas was very small, the manufacture of most material necessities was done in the home, and such trade as existed was carried on largely between the mother country and the colony, with raw materials moving to the homeland and the finished goods outward to the colonies. Indeed, had there been any great need for bridges, the colonial builders could hardly have satisfied it. The major coastal cities either lay upon or were dissected by waterways of great breadth and depth. We have only to recall that the Hudson was not bridged at New York until 1931, or the Mississippi at New Orleans until 1936, to realize how formidable the problem of large-scale bridge construction is. In the colonies there were no public bodies to raise taxes or issue bonds, and no private corporations with the means to invest money in toll structures. Yet bridges offer obvious advantages over ferries, and the colonists recognized their necessity wherever produce had to be moved from farm to town and the intervening streams were broad enough to make fording impossible but small enough to make ferrying a nuisance.

Most bridges in the colonies were constructed of logs or hewn timbers on piling, although smaller structures of rubble masonry held in place mainly by friction appear to have been built in New England and Pennsylvania. Masonry bridges with dressed blocks laid up in mortar would have been prohibitive in cost. The pile-and-beam type of timber bridge was invented by Roman military engineers, from whom the Celtic and later the Anglo-Saxon people probably learned the art. The technique was developed well enough for the Saxons to bridge the Thames at London at the end of the tenth century. The factors governing bridge construction in the colonies continued to be major determinants until the Civil War. Wood was plentiful; it was durable and strong in both tension and compression and easily worked, and its properties were long familiar. The rapid growth of agriculture and commerce in the colonies, as well as in the republic, meant that it was most expedient to build bridges for subsequent replacement rather than permanence, or to build with the aim of strengthening and enlarging the structure in the near future.

The primitive bridge of the colonies was simply a felled trunk with the top hewn flat that was laid across a stream from bank to bank. Later two trunks, or a large trunk split in two, were laid side by side. The earliest intermediate supports were crude pile frames of two trunks capped by a cross piece to take the ends of the logs or planks that formed the deck. The later built-up structure of hewn timbers consisted of the following elements: transverse rows of piles driven into the stream bed, the number of rows depending on the length of the bridge, with longitudinal girders spanning from one group of piles to the next; light transverse beams laid across the girders; and a plank deck nailed to the beams. The separate piles of any one group were joined by a transverse beam or cap piece laid across the tops. Such a group of connected piles is called a bent, and the longitudinal girders are known as stringers. Methods of joining appropriate to building frames could be used in bridges, but since iron spikes were stronger and easier to place, they eventually came to replace the wooden pegs and tenons. Span length in the colonial bridge seldom exceeded 20 feet and was usually shorter. The easier construction of short spans, however, was offset by the higher cost of driving a greater number of piles, the most laborious and expensive operation in building pile-and-beam bridges.

For bridges of some weight, shore supports were a necessity, since stream banks were generally too soft to sustain even a moderate wagon traffic. Abutments were usually built up of logs laid horizontally in a three-sided crib with notched joints to make stable corners. The wide interstices were filled with earth, clay, or field stones. Abutments of loose rubble were sometimes used, but these were exposed to the undermining action of currents. For bridges of more than one span in which piers took the place of piles, the piers were generally log cribs filled with loose stones. Foundations presented the most difficult problem to the colonial builder, since he was not equipped to build in cofferdams from bedrock or hardpan. The simplest technique was to stabilize the sediments of the stream bed by spreading stones over the area under the pier. A more reliable foundation was the weighted raft—a grillage of timbers laid in two layers at right angles and sunk in place by boulders. The weight of the stone-filled pier then held the raft in place.

The earliest bridges over major streams were built in the Boston area. The first on record was the Cradock Bridge, constructed about 1630 in the pile-and-beam form to span the Mystic River at Medford, Massachusetts. One of the longest was the Great Bridge, built over the Charles River at Cambridge in 1662. The stringers were set at variable spans, from 15 to 20 feet in length, and were carried on piers in the form of stone-filled cribs. The plank deck was spiked to transverse beams. The Great Bridge survived until nearly 1800, and up to 1786 it was the only span between Boston and the communities on the north bank of the Charles.

Samuel Sewall of New England was the major bridge builder of the eighteenth century. The first of his structures, and the oldest for which the builder's drawings survive, was the pile-and-beam bridge over the York River at York, Maine. The span was completed in 1761 and remained in continuous use with repairs until 1934. It was 270 feet long, 25 feet wide, and its deck was supported by 13 braced bents of four piles each. The bents were assembled on land, floated out to position, and sunk for some distance through the soft mud of the bed to a stable footing. The largest of Sewall's bridges was completed in 1786 across the Charles at Boston. Its 1,503-foot deck rested on 75 bents, with the usual maximum span of about 20 feet. The piles were hand-driven into the bed by means of a heavy block raised by tackle and allowed to fall by gravity.

The pontoon bridge was the only other timber form to be built in the colonies. Two such bridges were associated with the early campaigns of the Revolutionary War, but few details of their construction remain. The first was built over the Schuylkill River at Philadelphia in 1776 to move troops in Washington's New Jersey campaign. It was built under the direction of Israel Putnam, the general in command of the Continental Army at Philadelphia. The bridge has been described as constructed like a graving dock for ship repairs. A second structure was built by the British Army in 1777 over the same stream near Gray's Ferry but was soon replaced by a pile-and-log span.

The simple beam structure was not superseded until the last decade of the eighteenth century, when Timothy Palmer revived the timber truss bridge after two centuries of neglect. A transitional form which was an elaboration of colonial antecedents was the bridge built by Enoch Hale over the Connecticut River at Bellows Falls, Vermont, in 1785 (Fig. 7). A two-span structure with a total length of 368 feet, it has been regarded as the first truly framed bridge in the United States. It was a primitive variation of a type that first appeared in the early second century A.D., when Apollodorus of Damascus built the great wooden span over the Danube River for the Roman emperor Trajan. The superstructure of Hale's bridge rested on abutment walls of rubble and on a braced timber bent at the center. The deck planking and beams were carried on heavy braced timbers arranged in a trapezoidal archlike form, with four such arches set in parallel planes for each span. What we know of the construction suggests that the deck beams, arched stringers, and center bent constituted a framed structure that bears some resemblance to a truss in action. Hale's bridge was demolished sometime before 1803.

Although the truss bridge came to be a necessity for road and rail traffic over major streams, the pile-and-beam form with braced bents never entirely disappeared. It survives to this day for very short spans on railroad branch lines and rural roads. One may say as much, as a matter of fact, for timber construction in general. The ingenuity shown by the colonial carpenter was constantly improved by the builders of the early republic and adapted to a steadily widening range of structural needs as the industrial base of the new nation expanded. In American building the great period of timber framing was to come with the nineteenth century, but certain aspects of the technique were to become permanent features of the structural arts. For all the native ingenuity, however, American dependence on Europe for ideas, fundamental inventions, scientific theory, and mathematical techniques was also to remain a permanent characteristic of the construction industry. The European builder has always been more scientifically minded, more aware of his dependence on scientific theory and hence more flexible in his adoption of new ideas. His American counterpart has seldom freed himself from a narrowly pragmatic point of view, fostered by and at the same time inhibited by the corporate demand for immediate financial return.

Chapter Two | Masonry Construction

The proudest achievements of colonial architecture were constructed with brick or stone bearing walls carrying the beams, joists, and rafters of the floor and roof frames. Brick has always been the more useful building material for a number of reasons that are still important today but were compelling in the colonial period: the clays from which it is made are widely distributed and lie close to the surface of the ground; the process of manufacture before mechanization was a simple handicraft technique, far less expensive than the tedious and exacting techniques of quarrying and dressing stone; finally, brick walls are strong, durable, highly resistant to weathering, and more resistant than stone to the corrosive effect of smoke-laden air. Colonial brick-building, like the techniques of timber framing, was derived from medieval and Renaissance precedents. The construction of brick masonry was a well-developed art in the Middle Ages, useful for flooring, paving, walls, and vaults, and it had behind it a continuous history extending back to the third millenium B.C. The method of preparing brick from clay had scarcely changed over more than forty centuries: the process of digging, pulverizing, weathering, and kneading the clay and of forming the bricks and baking them in kilns was essentially the same in the American colonies as in ancient Mesopotamia.

Stone masonry is equally old and was carried to levels of complexity and refinement in the late Middle Ages that have never been surpassed. But to people with the modest economic resources of the English-speaking colonists, building in stone was likely to be very rare. Cobbles and loose fragments, available on the ground or in the beds of small streams, could be easily collected and placed in the wall by hand, but more refined forms involve the costly processes of quarrying, dressing, and transporting the stone and of laying the heavy blocks in carefully leveled beds of mortar. Domed and vaulted construction requires elaborate systems of scaffolding and centering which would have taxed the ingenuity of the carpenter as well as that of the mason, with the consequence that it was simply prohibitive in cost. Yet the American colonies were blessed with extensive resources of building stone; and where the money and skill were available, stone masonry became fairly common in certain regions by the mid-eighteenth century.

As in the case of the carpenter, the tools of the quarryman and the mason were inherited from the artisans of the Middle Ages. They included the iron-shod wooden shovel, the pick, wedge, crowbar, and maul for breaking the stone, and assorted poles, hooks, and beaks for disengaging the pieces from the natural mass. The hammer ax, broaching ax, chisel, saw, drill, and mallet were used for shaping and dressing the rough blocks. The lime required for mortar to lay up brick and stone walls was often derived by the colonists from the shells of molluscs, but they also followed the medieval practice of burning limestone or chalk. None of the limes so produced contains the alumina and silica necessary for the formation of hydraulic cement, which will set in water as well as in air. Limited to the non-hydraulic varieties, the colonial builders were barred from rigid masonry construction under water.

The most primitive forms of protomasonry construction were rammed-earth walls and walls of loose rubble piled in a circular or elliptical plan to increase the stability of the structure. There is little evidence of this kind of building, which was apparently confined to Virginia and Massachusetts Bay in the early years of those colonies. Building in brick began shortly after the initial essays in timber framing. The town of Henrico, Virginia, near Jamestown, had several two-story framed houses with first stories of brick as early as 1611. The second church at Henrico rested on brick foundations, which had been laid in 1615. Brick kilns were built at New Amsterdam in 1628, and a brickyard, possibly with associated kilns, was established at Salem in the following year.

The most rapid development of brick construction came in Virginia, where aristocratic taste demanded and wealth made possible the elegant plantation manor. Moreover, the settlers of Virginia came from all parts of England, including regions where brick construction was common, whereas the New Englanders came chiefly from the eastern and southeastern counties, where it was rare. The brick manorial house became common on the Virginia plantations in the latter half of the seventeenth century, by which time native brick was cheaper than the English varieties and was exported to Bermuda in return for limestone. The usual pattern of brickwork in the seventeenth century was English bond, the more elaborate Flemish bond appearing after 1700 (Fig. 8). The simple rectangular enclosure of the Virginia house was interrupted by the extremely broad multiple-flue chimney, which was wide enough at the base to embrace the fireplace. The oldest brick house, built by Adam Thoroughgood in Princess Anne County between 1636 and 1640, included a chimney that occupied half the width of the house up to the ridge line of the roof.

The progress of the brick mason's technical skill and the developing taste that called it forth are best exemplified by Bacon's Castle, a monumental house built about 1655 in Surrey County, Virginia. It stands in restored condition, the first American house with a cruciform plan and the first fully mature work in the Tudor Gothic style. The distinguishing features of the brick structure are the chimneys in the form of square-section shafts set with their diagonals in line and the high gables of Flemish design. The wooden floors are carried on joists spanning between heavy girts set into the masonry walls.

Among public buildings the most representative work springing from the northern Gothic tradition of brick masonry is the Newport Parish Church at Smithfield, Virginia, completed in 1682. The heavy brick walls of this modest building are two feet thick and are further strengthened by the addition of three stepped buttresses on each side which extend another two feet beyond the outer wall plane. Each bay defined by adjacent buttresses contains a single window opening surmounted by a brick arch in the pointed Gothic form. The front elevation of the church is dominated by a massive brick tower, square in plan and 20 feet on the side, with brick walls equal in thickness to those of the main enclosure. The original structural elements of the interior have disappeared, but the presence of the buttresses suggests a primary roof frame of heavy arched rafters that exerted a horizontal thrust at their lower ends. The gallery at the rear end of the nave was probably supported on beams spanning between the rear brick wall and a deep transverse girder.

The Dutch, who were the most skillful brick masons of Europe, brought their art to New Amsterdam shortly after the colony was founded. They were the first to build with glazed brick and tile and the first to make brick in a variety of colors, such as red, salmon pink, orange, yellow, and purple, achieved by varying the temperature and the time of baking in the kilns. The presence of various salts in the clay also affected the color and was the source of the glazed surface. The Dutch bricks were usually narrower than the English, about 1½ inches thick as against 2½ inches, the latter being standard today. By using a variety of bonds—common, English, Flemish, and Dutch-cross—along with contrasting colors and glazes, they built walls in elaborate patterns, the lively geometric effect enriched by colored roof tiles and stepped gables. Interior timber construction, however, was usually simpler than the New England system: plank floors rested directly on a closely ranked series of heavy beams spanning from wall to wall.

Many of the masterpieces of colonial Georgian architecture were derived without essential structural change in the masonry work from the traditions of the seventeenth century. Brick became dominant for the exterior walls of all major buildings; for the interior structure, variations and extensions of heavy braced framing continued to be used to support floors, roofs, towers, and spires. The only new materials added to the eighteenth century building were paint and lime plaster. Among the large public buildings, Christ Church in Philadelphia is as much a triumph of the mason's art as it is the carpenter's. The Flemish bond of the brick walls is distinguished by the systematic use of glazed headers throughout the wall area, a decorative device that provides a good example of making a virtue out of an accidental circumstance. (A header is a brick laid with its end face in the outer wall plane; a stretcher is one laid in the conventional manner, with the long narrow face in the outer plane.) The black glaze comes from the coincidental presence of two factors: one, bricks made of clay containing a relatively high proportion of potassium oxide (about 3 per cent), and the other, the fusion of this oxide into a glaze at the ends of bricks closest to the kiln fire. The eighteenth-century mason selected these glazed ends for use as headers, thus intensifying the pattern of the bond by introducing a rhythmic variation in color and texture. The nave walls of Christ Church are strongly articulated by means of the projecting courses and the piers between the high arched windows. The high tower required walls four feet thick, in which the embrasures for the round-headed windows are splayed outward toward the inner wall surface to maximize the admission of light. The conical surface of these arched embrasures required specially molded brick even when laid in the conventional way, but in some of the arched openings the mason sought the ultimate expression of his virtuosity by setting the brick in a herringbone pattern.

The tower of St. Michael's Church in Charleston represents a more daring essay in masonry construction because of the change in cross-sectional shape. The tower is square in section up to an elevation a little above the ridge line of the nave roof, above which there are two octagonal stories, respectively a belfry and a clock enclosure. The builder was thus faced with the necessity of supporting the four diagonal faces of the octagon over the voids extending across the corners of the square. These faces are carried on corbeled brick arches known as squinches, which span across the corners of the square and support the brick masonry above them. The squinches do not appear on the exterior but are visible on the interior at the base of the first octagonal story. The smaller octagon of the second story is supported by corbeling the interior faces of the brickwork inward at the top of the belfry. The height and slender proportions of the tower required a wall thickness at the base of 4 feet 9 inches.

Nearly all masonry work in the English colonies belonged to the categories of straight bearing-wall or pier construction, with arches largely confined to those parts of the wall surmounting arched openings for windows and entranceways. Full domes and vaults were never used, chiefly because of the expense of building the necessary centering and falsework. The top profiles of window openings were usually semicircular in form, following the classical precedent, but in some buildings of the seventeenth century other shapes appeared, ranging from the flat-arch lintels of Fort Crailo, built by the Dutch at Rensselaer, New York, in 1642, to the Gothic arches of the Newport Parish Church.

Special forms of structural arches were employed in Independence Hall in connection with the heating and smoke dissipation systems. Since the fireplaces on the second floor could not have hearths resting on the timber frame, each was carried on a half-arch (or more strictly, a half-vault) springing from the brick end wall and supported at its free end by wedge-shaped pieces fixed to blocks extending on either side of one of the floor girders. A larger and more daring form of arch construction is associated with the flue and chimney structure. The four-chimney group at each end of the hall rests on a massive brick arch set against and parallel to the end wall, thus bringing the load of the four chimneys down to two pierlike masses projecting from the wall on either side of the two fireplaces located at each end of the building. The chimney arch in the east wall is semicircular, whereas that in the west is a flattened segmental form made up of three circular segments drawn about separate centers.

Stone masonry, as we have seen, was rare in the colonies beside brick and timber construction. It was concentrated mainly in the Hudson valley of New York, northern New Jersey, and eastern Pennsylvania. In New England the extensive granite outcroppings lie in enormous masses, discouragingly expensive to quarry with the colonist's limited means, and in the Piedmont and along the coast the soil is composed chiefly of marine and alluvial sediments. The original form of stone masonry, first confined to chimneys and later used for walls, was an irregular rubble of stone fragments found lying naturally on the ground, but sometimes broken from ledges or split off from weathered masses. Up to the mid-seventeenth century the practice was to lay the pieces in clay bound with straw, but mortar made from sand and oyster-shell lime soon became standard for durable building. There was a steady improvement in masonry techniques throughout the century: stones were carefully selected for uniformity of size and shape and laid roughly to course; eventually the stones were dressed and laid with even joints to form what is called ashlar masonry. In some areas, a special kind of stone construction known as cobble masonry was used, in which the wall is made of small rounded boulders set in a thick matrix of mortar. Cobble houses became common in central New York, chiefly in the Mohawk valley, where there was a plentiful supply of these glacially distributed stones. The walls of all stone buildings tended to be very thick, running from 18 to 36 inches even for ordinary residences.

Stone houses were common in the Dutch building of the New York area, and a few still survive in the city itself. A good example of irregular masonry is the Billopp house at Tottenville, Staten Island, built near the end of the seventeenth century. An early classic with walls of carefully selected regular blocks is the Ten Broek house, built between 1676 and 1695 at Kingston, New York. Stone masonry seems to have encouraged experiments in form, especially among English and German builders in the Philadelphia area. The house built by Sir William Keith, governor of Pennsylvania, for example, is an extremely long, high, narrow block capped by a gambrel roof. Located in Montgomery County, the Keith house was constructed from 1721 to 1722. The Georg Mueller house, built at Milbach, Pennsylvania, in 1752, spreads out in a number of wings, its rubble walls contrasting sharply with the smooth blocks of the quoins.

The arch of stone masonry was even rarer than its brick counterpart because of the added cost of dressing stones, however crudely, into the wedgelike shapes of the arch voussoirs. What is very likely the earliest use of the stone arch appears in the most puzzling of all colonial structures, namely, the Old Stone Mill at Newport, Rhode Island, built sometime before 1677. The mill is a squat cylindrical tower of irregular stone fragments carried at its base on eight cylindrical piers joined by crudely made semicircular arches in which the pieces of stone are laid roughly as voussoirs. The outside diameter of the tower is about 23 feet, and the wall thickness and pier diameter are 2 feet 6 inches. Who erected this strange tower and to what purpose have never been determined. One romantic theory is that it was built by Norsemen who were thought to have explored the Newport region in the fourteenth century. The only reliable evidence indicates that the tower was owned and used as a mill by Governor Benedict Arnold, whose will of 1677 contains the first reference to it. The cylindrical mill itself was not uncommon, having been built by the French in the form of solid towers without arcades along the St. Lawrence River and at St. Louis. The stone arch was ordinarily used in connection with window and door openings, in a few cases in the Gothic form, as in Sleepy Hollow Church at Tarrytown, New York (1699), but always during the eighteenth century in the Roman semicircular form, as in the great stone house known as Mt. Airy in Richmond County, Virginia (1758–62).

Arch bridges of stone were extremely rare in the colonies, and reliable records are non-existent. There is scarcely any evidence for the construction of stone bridges in the seventeenth century, and there is little to suggest the exact form of those built in the eighteenth. The first such structure was the multiple-arch bridge built in 1693 to carry Frankford Avenue in Philadelphia over a local stream. Possibly the second was the stone bridge built in 1740 to carry Third Street over Dock Creek in Philadelphia. This was undoubtedly an arch bridge, although the available descriptions do not specify the structural character. A bridge of unspecified material was known to have been built in 1720 to carry Second Street over Dock Creek, and since it cost half again as much as the Third Street bridge, it may also have been of masonry construction. We may conclude that there was steady progress in the art during the late colonial period because a nine-span stone arch bridge was constructed at Lancaster, Pennsylvania, at the end of the century. Construction in stone masonry continued to flourish in the first half of the nineteenth century, but thereafter its role was progressively superseded by iron and concrete.

The great works of masonry construction were built in the Spanish colonies that once stretched without break from Florida to California. The Spaniards, who came to find precious metals and to convert the Indians, established their permanent settlements on feudal and hierarchical lines, with the consequence that the wealth of church and state was available for the construction of missions, fortifications, and government buildings. The richness and complexity of Spanish Baroque architecture could thus be duplicated in the New World in forms and on a scale impossible to the bourgeois economy of the English colonies and antipathetic to their Protestant spirit. Whereas the building inheritance of the English colonists were initially medieval, that of the Spanish domains was derived from neo-Renaissance precedents at the start.

The first permanent settlement of the Spanish in the area of the United States was founded in 1565 at St. Augustine, Florida. Here a century later the colonial governor built the first great work of military architecture in North America and one of the early masterpieces of masonry construction. The Castillo de San Marcos, standing today as a national monument, was begun in 1656, largely completed between 1672 and 1675, and extended in 1756. The Castillo is typical of the bastioned fortification developed for artillery combat in the period from the fourteenth to the seventeenth century and given its ultimate form by the French military engineer Sebastien de Vauban (1633–1707). The walls of the bastion, or the curtains, are built up of dressed blocks of coquina limestone laid in mortar made from sand and oyster-shell lime. The stone is a soft white limestone composed of broken shells and coral cemented under pressure into durable strata. The outer curtains of the fort are 25 feet high, 12 feet thick at the base, and 7 feet at the top. The coursed masonry constitutes the outer covering of a rubble core, a composite masonry form that was to survive in the construction of large dams until about 1910. The most remarkable feature of the Castillo is the presence in one of the cells of a concrete floor made from a mixture of sand, oyster-lime cement, and a crude aggregate of broken shells, the surface of which was rubbed smooth and finished with linseed oil. This undoubtedly represents the first use of concrete in America.

The fully developed Baroque architecture of Spain appeared in the large mission churches that were built from Texas to California in the eighteenth century. The earliest of the major works is the Mission San Jose y San Miguel de Aguayo, constructed at San Antonio, Texas, 1720–31. The solid bearing walls are built up of rubble masonry laid in lime mortar. The bearing stone is tufa limestone, a porous, calcareous stone somewhat like travertine in its appearance and physical properties, and the ornamental details are carved in brown sandstone. The exterior surface is covered with stucco, which hides the irregular jointing under a smooth coating. The forward part of the nave is roofed by three groin vaults in linear succession, the arched groins springing from exterior wall buttresses that absorb the horizontal thrust of the arches. The rear bay of the nave is covered by a hemispherical dome rising to a height at the crown of 60 feet above the floor. The vaults and the dome collapsed in 1873 and were restored in concrete. The neighboring Mission Nuestra Señora de la Purísima Concepción, constructed in 1736–55, has been in continuous use without alteration to its original fabric. Along the south wall of the San Jose nave is a low baptistry roofed by three domes on pendentives (that is, spherical triangles contracting downward), which provide a transitional bearing from the circular periphery of the dome to the square outline of the bay. Behind the nave is a sanctuary covered by a barrel vault. The tufa blocks of the vaults and domes were laid up on wooden centerings and covered with stucco after the mortar had set. No masonry work in the English colonies remotely approached the technical complexity of the Mission San Jose. Neglect of this structure and others like it following the separation of Texas from Mexico led to the partial collapse of the San Jose church in 1868 and again in 1873, but in 1933 it was restored approximately to its original condition.

The leading structural archievement among Spanish ecclesiastical buildings is the Church of San Xavier del Bac, constructed at Tucson, Arizona, between 1784 and 1797. The walls throughout are built up of kiln-baked brick laid in lime mortar and covered with white stucco. The most remarkable features of the church are the buttresses that serve to maintain the stability of the relatively narrow octagonal towers above the flanking wings. A substantial part of the tower load is carried to the corner piers of the wings by flying buttresses whose surfaces are arched below and developed into scrolls above. The nave, transept wings, and apse are covered by low domes of brick; a high dome on an octagonal drum roofs the crossing. This drum is carried to the four corner piers at the junctions of the nave and transept walls by means of inverted pyramidal blocks corbeled out from the pier faces (these may be regarded as pendentives with flat triangular faces). The extremely thick walls of the church sustain the outward horizontal thrust of the domes, and the opposing thrusts along the inner segments of the peripheries of contiguous domes cancel each other. These technical details indicate how nicely the builder calculated the distribution of material for structural purposes in the San Xavier Church.

The less pretentious ecclesiastical buildings were combinations of timber and masonry construction. The unique example of this kind is the Church of San Carlos Borromeo at Carmel, California, erected between 1793 and 1797. The nave is roofed by a vault in the form of a parabolic cylinder with a clear span of 29 feet. The vault is constructed of wooden planks spanning longitudinally between three transverse arches of stone. These arches spring from shallow stone buttresses projecting inward from the inward curving surfaces of the masonry walls, the curving profile introduced to maintain a continuous surface from the crown of the vault arch to the ground.

For the smaller churches and the secular buildings of New Mexico the Spanish colonists often used adobe construction, the only form of permanent building derived to some extent from native traditions of the New World. The agricultural Indians of New Mexico built large communal structures called pueblos for dwelling, storage, and defensive purposes as early as the tenth century. They were constructed in terraces, four or five stories in height, with the lower floor devoted to the storage of food. The pueblos were sometimes built of broken slabs of stone but more often after the sixteenth century of a kind of brick called adobe, after the Spanish word for brick, which in turn was derived from the Arabic *atob*. In this form of construction a solid bearing wall was built up of wide, sun-dried clay bricks or of sun-dried clay molded into flat layers. The surfaces were then smoothed with adobe plaster, a thin wet clay mixed with gypsum (calcium sulfate). For the roof construction closely spaced beams in the form of round logs (called *vigas*) were laid transversely on the tops of the walls. Thin branches, sticks, or reeds, laid in a dense mass over the logs, carried a thick blanket of clay that made a durable roof slightly pitched toward drain spouts outside the walls.

The secular buildings and small churches were constructed by the Indians under Spanish direction but largely according to native traditions of workmanship. The chief innovation introduced by the Spaniards for high-walled churches was to build one wall thicker than the other to serve as a working platform and as a fulcrum for lifting the heavy log beams into position. The most impressive example of this technique is the Church of San Estevan at Acoma, New Mexico, completed in 1642. The long side walls of the nave are of unequal thickness, one 7 feet at the base and the other about 5 feet, the thickness of both reduced about 30 inches between base and top. Indian women kneaded the clay and formed it in molds into sun-dried bricks each weighing 50 to 60 pounds, or about as large as one man could conveniently handle. At the top of the inner surface of the wall, carved brackets were set into the clay to provide bearing surfaces for the *vigas*. Some of these were left round, others were hand-hewn to square section, but all were long enough to span the 30-foot clear width of the nave. Spaced about 3 feet on centers, they carried hand-hewn planks and cedar poles set in a close herringbone pattern. A heavy blanket of clay 6 to 12 inches thick was then laid over this base to provide the finished roof. The whole enclosure of church and associated *convento* is a masterpiece of primitive building, perhaps our foremost example of Indian structural techniques surviving in a permanent structure.

The climax of the Spanish building tradition in North America proved also to be its conclusion. The Cabildo at New Orleans, constructed between 1794 and 1797, was built to house the offices of the Spanish government of the city, which lasted from 1763 to 1801. The building is one of a pair of identical structures flanking St. Louis Cathedral, the other being the Presbytère, begun by the Spanish in 1794 as a home for parish priests and completed under United States sovereignty in 1813. The two are architecturally and structurally almost identical, differing only in minor details of plan. The Presbytère was extensively altered in the interior during a recent renovation, but the Cabildo stands very nearly as originally constructed (Fig. 9). Roughly L-shaped in plan, it stands three stories high under a mansard roof. The foundations form the one crudely primitive part of the construction and as a consequence have been a source of trouble throughout the history of both buildings. Shallow trenches were dug around the perimeter, a clay blanket spread over the base of the cutting, and cypress planks laid over the clay up to about six inches below grade level. Spread footings of brick for the piers and walls were built up directly on the planks. Subsequent lowering of the water table by drainage exposed the wood to rot, with the result that the walls and piers settled irregularly, cracking the stone flooring in places and causing the St. Ann Street wall of the Presbytère to lean 11 inches out of plumb.

The exterior structure of each building is brick covered with stucco that has been scored to imitate stone masonry. The façade is divided into ten piers, which carry an arcade at the first story, but the walls of the side and rear elevations are solid bearing masonry. The interior structure is entirely red cypress timber except for an arch and two piers which help to support the second floor over the area of the largest interior enclosure. The floors are heavy planking carried on the familiar framing system of beams and joists. The mansard roof at the third story and the hipped roof above it, both covered with thick slate, required an elaborate system of framing that has certain unique features among colonial buildings. Although the present mansards represent mid-nineteenth-century alterations to the original roof structure, they embody the structural system that had been used for trussed mansard roofs since the early sixteenth century. The hipped roof rests on a system of purlins and rafters that are carried in turn on a series of trusses spanning transversely between two rows of heavy sloping posts that bear on the top of the brick walls, the angle of inclination being the same as that of the steeply sloping planes of the mansard. The rafters of this roof stand independent of the posts except for a short horizontal piece joining post and rafter near their bottom ends. All connections are secured by means of tapered wood pins inserted through mortise-and-tenon joints.

An octagonal drum and slate-roofed cupola located above the center point of the ridge line offered special difficulties in framing. The drum is framed with eight timber posts joined by double-diagonal bracing in the vertical planes. The posts, which rest on a braced octagonal frame set between the bottom chords of the two center trusses, carry a ring girder the segments of which were carefully cut to fit the circular shape of the ring. This ring supports the eight rafters of the cupola roof, the rafters having been shaped to fit the reverse curve of the roof section. A massive center post within the drum provides the main support for a helical stairway.

In spite of the advanced stage to which the Spanish builders carried masonry and timber construction in their colonies, their techniques left little influence on the development of North American building outside the southwestern United States, where Hispanic forms and techniques remained common to the twentieth century. Spanish rule and its associated cultural elements were rapidly obliterated by war and the westward movement of English-speaking pioneers in the first half of the nineteenth century. The institutions and the social character of the United States were formed to an overwhelming degree by North European peoples, who felt little sympathy for Mediterranean and Catholic cultures, and as we might expect, architectural and structural techniques tended to follow social development. Some of the finest achievements of Spanish building art survive to this day, but only as historical monuments preserved by the federal government or the Roman Catholic church. We admire them now as works that belong entirely to the past, for they left no continuing legacy in the rising republic.

Part Two

The Agricultural Republic: Revolution to Civil War

Chapter Three | Timber-Framed Buildings

The separation of the colonies from England following the Revolutionary War brought drastic alterations in the entire pattern of American economic development. The imperialist practice of extracting raw materials from the colonies, manufacturing the finished product in the mother country, and selling it in turn in the colonial market was superseded by a more balanced internal commerce. There was some preparation in New England for the rise of a capitalist industry, but on a scale wholly inadequate for the needs of the growing republic. The textile, machinery, and tool industries, for example, are largely post-Revolutionary developments. With industrial expansion came the growth of financial and administrative institutions and the extension of transportation arteries. The first canal dates from the 1790's, but the great period of canal building did not come until the early decades of the following century. The first railroad was built in 1830, and by 1850 it had established itself as the primary form of transportation. Although the United States as a whole was predominantly agricultural up to the Civil War, economic power was steadily being concentrated in the industrial and financial centers of Boston, New York, and Philadelphia. The major determinants in shaping the building techniques of the early republic soon came to be industry, finance, and the railroad, with the last destined to become the most potent factor. The monumental buildings of state and federal governments were ultimately determined by the same factors, since economic expansion was accompanied by a parallel expansion in the role of government. The skylines of American cities continued to be dominated by church steeples, but as an index to the character of urban building, this was often an illusion: in rail and manufacturing centers, the secular buildings bulked far larger in number, size, and complexity.

Wood was the obvious choice for the overwhelming majority of buildings and bridges in the early history of the republic. Timber was plentiful, and the enormous areas of forest land meant that transportation from source to town was an inexpensive short haul. Wood possesses a wide range of desirable physical properties: it is strong in both tension and compression and hence can resist bending forces as well as direct compressive loads; it is durable if protected from rot, even under conditions of extreme weathering; and it can readily be shaped into a great variety of structural members by simple hand-and-tool techniques. In a nation where land seemed to be available in unlimited quantities and farming always offered a means of livelihood, manufacturers were always plagued by a shortage of urban labor. In America, in contrast to Europe, materials have always been cheap and labor costly. The main urban centers of the young nation possessed the means for large-scale building in masonry, but they could seldom command the economic resources. The skills of the carpenter, on the other hand, were nearly universal among able-bodied men who worked with their hands. The introduction of the nail- and spike-cutting machines after 1790 and of the power-driven circular saw in 1814 greatly increased the production of boards and heavy timbers. But beyond these mechanical devices, traditional handicraft techniques were sufficient to build the largest mills and bridges. The pragmatic, rule-of-thumb approach of the craft tradition provided the carpenter with all the science he needed.

The heavy New England frame of posts, girders, beams, and joists could be elaborated and enlarged to provide the structural basis for a great diversity of buildings—mills, warehouses, railroad stations and roundhouses, stores, dwellings, and office blocks. Where necessary, it could be braced either horizontally in the floor framing system or vertically between posts. Timber frames were usually limited to an area in plan of 50 × 150 feet and to a height of three stories, but in a few cases an attic story might be added to the three working floors. The adaptation of the heavy timber frame to industrial building came with the construction of the Slater Mill in 1793, the first textile mill in the United States. Still standing at its original location on the Blackstone River at Pawtucket, Rhode Island, the building is three stories high, with a shallow attic under a gable roof. All floor, roof, and wind loads are carried by an unbraced frame of massive columns and beams roughly square in section. The whole structural system is simply an expansion of the basic New England frame to the size necessary to support a three-story mill. A stone foundation carries the timber sills into which the wall columns are framed; the intermediate columns rest on separate stone footings. Girders span longitudinally over the tops of the columns at each story, and the floor beams and roof rafters are framed into them. Joists extending between the floor beams carry the plank floor, and studs set between the wall columns support the sheathing and clapboard siding that constitute the curtain wall.

The range of industrial uses for all-timber construction was constantly widened in the early years of the republic. The most remarkable examples of the carpenter's daring and versatility were the short-lived "ship houses" built at the Philadelphia Navy Yard between 1821 and 1823 to provide sheltered enclosures in which to construct the hulls of naval vessels. If there was little art expended on these ungainly structures, their odd shape and astonishing size taxed the carpenter's ingenuity. They were immense wooden sheds, entirely open inside and at the ends and hence might be regarded as the timber equivalent of the modern aircraft hangar. The designer, either Joshua Humphreys, a naval architect, or Philip Justus, a Philadelphia builder, may have created an independent invention in the ship houses, but he may just as likely have derived the essential idea from the boat slip constructed in 1772 at the Devonport Naval Dockyard in England. The first of the two houses measured 74 × 210 feet in plan and 80 feet in height; the second had corresponding dimensions of 84 × 210 × 103 feet. The structural system is not known exactly, but it can be inferred from surviving prints. The walls were built up of inward-leaning posts made up of several lengths, all tied together longitudinally by horizontal beams set in the plane of the posts and covered with siding. The gable roof was made of planking fixed to rafters that extended from the ridge beam to the plates at the top of the two rows of posts. Three levels of windows in the siding, along with the open ends, admitted light to the interior. There is no evidence of bracing, but without it these huge enclosures would have been vulnerable even to moderate wind. The ship houses did not survive long and they were isolated phenomena in structural evolution, but while they lasted they must have provided impressive demonstrations of the potentialities of timber framing.

The system of column-and-girder construction in timber served admirably for early industrial buildings, but it possessed one great defect in its vulnerability to fire. The first step in reducing this danger was the substitution of stone or brick for the plank and clapboard walls, a transition that occurred around 1810. The original Georgia Mill at Smithfield, Rhode Island, built in 1812, was one of the first with exterior bearing walls of stone and interior framing of timber columns and beams. Zachariah Allen, the foremost entrepreneur of the Rhode Island textile industry, acquired the Georgia Mill and enlarged it to a length of 250 feet between 1853 and 1854. With the addition of bracing and more rigid connections, its masonry and timber structure proved to be a sufficiently reliable form of construction to survive into the twentieth century for smaller factories and warehouses.

Although the expanded New England frame worked well enough for mills and other industrial buildings, it was heavy and awkward for houses, barns, and other small enclosures and often difficult to construct with the ruder skills and tools of the frontier settlements. These defects were overcome with the invention of the balloon frame, a widely useful innovation in structural techniques that was the first of Chicago's revolutionary contributions to the building arts (Fig. 10). The inventor was very likely George W. Snow, a local surveyor, but the first convincing demonstration of its possibilities was St. Mary's Church, designed by Augustine D. Taylor, a carpenter from Hartford, Connecticut, and erected in Chicago in 1833, about a year after Snow developed the technique. Starting with the familiar New England frame, Taylor removed all the heavy members, such as girts and posts, and reduced it to a framework of studs and joists, which could be assembled easily by hand with nothing more than a hammer, saw, and spikes.

In the developed balloon frame, a sill is laid directly on a foundation, originally made of upright logs, stone, or brick but now of concrete, and the studs are nailed into the sill usually at a spacing of 16 inches on centers. A plate nailed to the top of the studs carries the roof rafters, and these in turn support the purlins and roof sheathing. If there is a second floor, the plate or ribbon carries another row of studs and the floor joists. The floor boards are nailed directly to the joists. The boards of the wall sheathing are nailed to the studs, and siding is laid over the sheathing for an external covering. Today the undersheathing is nearly always a synthetic insulating material, and the outer covering is more often a brick veneer than wood siding. The roof sheathing is covered with a waterproof or water-shedding material. The framing members are mostly 2 × 4-inch pieces easily lifted and assembled by hand; the joists, however, are generally larger members because they carry greater loads over longer spans. The 2 × 4's may be doubled or tripled at places where internal and wind loads are concentrated, as at the edges of openings and the corners of the frame. Diagonal pieces are frequently introduced at the corners for wind bracing.

The ease and rapidity with which the balloon frame could be erected by a single carpenter was in part the result of the mass-production of nails, a manufacturing technique perhaps as revolutionary as the structural invention itself. The transition from the colonial hand-wrought nail to the machine-cut variety occurred in the period between 1790 and 1830. Many factories for the manufacture of cut nails were established in the industrial East, with the result that by the turn of the century the United States had surpassed England in the volume of production. The first nail-cutting mills were operated by hand power, but this stage soon gave way to operation by water power and eventually by steam. Numerous patents were granted during the first thirty years of the century, but since most of the records were destroyed in the Patent Office fire of 1836, the improvements in nail-cutting machines are conjectural and largely anonymous. There is sufficient evidence to show that the phases in the evolution of the process were marked by changes in the methods of shearing the bars and forming the heads. All carpentry involving the driving of nails and spikes depended on the supply of the machine-cut forms almost until the time of the Civil War, when the modern wire nail was introduced into the United States.

The balloon frame with its nailed connections offered such obvious advantages in economy and efficiency of construction and was so nicely adapted to the building needs of farms and small towns that it quickly spread throughout the Middle West and later over the nation as a whole. The great majority of suburban single-family dwellings built today are balloon-frame structures covered with a brick or stone veneer. Shortly before the Civil War two structural developments of great economic and social importance grew out of the new technique. One was the cruciform plan, in which four wings spreading out from a central-core provided a three-way outside exposure for all the rooms. The other was the prefabrication of building elements, which may be regarded as the initial step in the industrialization of the building process. The idea had Renaissance antecedents, but its full implementation was the product of nineteenth century technical developments in Europe and America. In the early phases of the process, prefabricated units of timber frames, walls, and roofs were manufactured in a factory, packed with directions for assembly, and shipped to a railhead, where the builder picked up the parts, carried them by wagon to his destination, and assembled them by hand. By 1860 the manufacture and shipment of prefabricated units for houses, barns, and stores was a major business in Boston, New York, and Chicago. The Civil War provided further stimulus through the demand for prefabricated field hospitals and prefabricated shelters for military supplies.

Although the balloon frame and its heavier ancestor of posts and beams served admirably for many building requirements before the Civil War, they were limited in their applications and unsuited to the more stringent demands of mechanized industry and the railroad. The industrial builder, for example, faced the crucial problem of supporting heavy machinery in multistory fireproof factories, and his counterpart on the railroad had to span relatively wide enclosures without intermediate supports, as in stations and roundhouses. These needs were met in two different ways: by the addition of iron members as supplements to the timber frame, and by the development of the timber truss far beyond the simple forms used in colonial churches. (For the internal evolution of the timber-truss bridge and iron frame, see Pt. II, chaps. 4 and 6.)

Iron was introduced into American building construction in 1822 by William Strickland, but there is no evidence that it was used in mills before 1850, and even at that date it was present only in the form of auxiliary members. An early example of this composite construction was the Pemberton Mill, completed in 1853 at Lawrence, Massachusetts, a work whose subsequent fate provided a costly lesson to the builders of large industrial structures. The floor loads of the mill were divided between exterior brick walls and an internal frame of cast-iron columns and timber beams, joists, and flooring. The 16-inch deep beams were reinforced with wrought-iron tie rods on their under or tension surface. It was a structural system that was to remain common for smaller office, store, and industrial buildings throughout the century, although after the Civil War it was rapidly superseded in the large cities by all-iron construction. But the Pemberton Mill survived for little more than six years: on January 10, 1860, the entire building collapsed, killing about two hundred of the six hundred operators on duty at the time. The subsequent investigation soon uncovered the immediate causes, but much more had to be learned before this common disaster of the nineteenth century could be avoided. The Lawrence accident resulted from a variety of defects in the mill structure: walls unstable under lateral forces, insufficient depth of the timber beams, and columns cast with eccentric cores, leaving too little metal along one element of the cylinder. Mills like the Pemberton were built according to the pragmatic tradition, but the use of iron introduced unknown factors whose action could be understood only after a thorough scientific investigation of the material and its appropriate structural forms. And with fire an omnipresent threat, the day was soon to come when iron would be a fundamental necessity for industrial and commercial construction.

Techniques for spanning wide enclosures grew out of the truss framing of roofs. Where the span was modest, the familiar king-post truss served well enough, especially if its rigidity was increased by the addition of diagonal struts. The invention of the Howe and Pratt trusses around 1840 provided the builder with much stronger and more rigid forms, so that he could drastically increase the length of roof spans without adding intermediate columns (for descriptions of these trusses, see pp. 61, 96–97). Since the requirement for wide-span enclosures was paramount in the case of the railroad train shed, the history of truss framing in the nineteenth century outside of bridge construction is an integral part of the history of the railroad station. Originally the station building and the shed were incorporated in a single structure in which the gable roof on rafters or trusses covered the tracks as well as the waiting rooms and other facilities. The Boston and Lowell station at Lowell, Massachusetts (1835), was the prototype of all stations built according to this integrated plan.

As traffic increased, however, the track area expanded with it, requiring not only a greater roof area but also the separation of the train shed from the station building. The growth in over-all size was rapid; to keep pace with it the builders stretched the gable roof on trusses to extraordinary dimensions. By 1850, as in the case of the mills, iron rods were introduced into the truss as supplementary members. The first La Salle Street Station in Chicago, built in 1853 by the Northern Indiana and the Chicago and Rock Island railroads, had a gable shed spanning 116 feet that was supported on huge triangular Howe trusses with timber chords and diagonals and wrought-iron rods for the vertical tension members. Two years later the Illinois Central Railroad built one of the largest rail facilities of the time in its Inbound Freight House at Chicago (Fig. 11). The over-all length was originally 572 feet 6 inches and the clear interior span 90 feet. The roof is carried on triangular Pratt trusses of wrought iron and timber, the trusses in turn resting on the thick walls of irregular limestone masonry. Partially destroyed in the Chicago fire of 1871 and rebuilt the following year, the freight house still survives in its original structural character. Variations of the wood-and-iron truss continued to be built by the railroads until nearly the end of the century. A French importation, the Polonceau truss, was used by the Atlantic and Great Western Railroad to support the roof of its station and track area at Meadville, Pennsylvania (1862). In this truss, invented in 1837 by Camille and Antoine Remi Polonceau, the posts are set normal to the inclined top chords. Howe trusses carry the gable shed of Dearborn Station in Chicago (1883–85), perhaps the last metropolitan terminal to be built in this way. The presence of two flanking sheds at the sides of the central track area made it possible to limit the main gable to a span of only 65 feet. The Dearborn shed was a curiosity that seemed to belong to a remote and primitive past in the building arts, but it was not finally demolished until 1976.

Long before Dearborn Station was constructed, however, the

builders of railroad terminals began to see the restrictions imposed on them by the traditional forms of gable roof and triangular truss. If a track area wider than 100 feet was to be roofed in a single span, the vault supported on arched trusses offered a much more efficient structural system. The pioneer essays in arched timber construction were the forerunners of the great balloon sheds of iron and steel that characterized the metropolitan terminals in the last quarter of the nineteenth century. The first vaulted shed in the United States covered the track area in the Philadelphia station of the Philadelphia, Wilmington and Baltimore Railroad, built between 1851 and 1852 after the design of George A. Parker, the company's chief engineer. The vaulted roof of timber planking was carried on arched Howe trusses of wood that spanned the full width of 150 feet between the brick side walls. The horizontal thrust of the arch was sustained by a wrought-iron tie rod extending between the springing points and supported at its center by an iron hanger suspended from the crown of the arch. The P. W. and B. terminal was the first of three Broad Street stations built in Philadelphia by the Pennsylvania Railroad and its predecessors.

The largest work of vaulted construction in wood was the shed of Great Central Station in Chicago, designed by Otto H. Matz and built between 1855 and 1856 as the first of the Illinois Central's terminals in Chicago. The arched Howe trusses had a clear span of 166 feet, a dimension that made the shed the second widest roof in the world at the time. (The first was New Street Station, Birmingham, England [1854], with a span of 211 feet for the train shed.) Wrought-iron tie rods extended between the ends of the arches to relieve the masonry side walls of part of the arch thrust. Great Central Station fell before the Chicago fire fifteen years after its completion. A small shed embodying the same system of construction was built by the Central Vermont Railway between 1867 and 1868 for its station at St. Albans, Vermont (Fig. 12). The shed covered four tracks and platforms with a clear span of 88 feet. The St. Albans station survived for ninety-five years before its demolition in 1963, a long life that was once a common characteristic of New England rail structures.

What is very likely the most ambitious work of timber roof framing in the world serves religious rather than secular purposes. The huge domed building in Salt Lake City known as the Mormon Tabernacle, or more correctly, the Tabernacle of the Church of Jesus Christ of Latter-Day Saints, is unique in its structural character. It was designed (1862–63) and constructed (1863–68) primarily as a place of worship and general assembly for the conduct of church affairs. The idea for this monstrous turtleback structure came from the mind of Brigham Young only fifteen years after the Mormons first descended into the desert of Utah's central valley. He wanted a single enclosure without intermediate supports large enough to seat about seventy-five hundred people (the later addition of a balcony raised this capacity to nine thousand). The architect in charge of the program was William H. Folsom, the interior architect Truman O. Angell, and the designer and supervisor of the roof structure Henry Grow. For an auditorium of this size, any conventional roof was out of the question, and a domed or vaulted structure was the only alternative. It was in the supporting framework of this vast cover that Grow demonstrated his ingenuity and daring as a designer of timber structures. A bridge builder for various railroads in eastern Pennsylvania, he had come to Utah as a convert to the Church of Latter-Day Saints. He built several bridges in Utah over the Weber and Jordan rivers in which the decks were carried by lattice trusses of timber, a form originally invented by the architect Ithiel Town (see pp. 58–59). These durable spans made a favorable impression on Young, who selected Grow as the designer of the tabernacle roof on this basis. The structural system he employed represents the first and the most extensive application of the lattice truss to arched roof framing.

Grow's problem was to build a domelike cover for an oval-shaped auditorium with over-all dimensions in plan of 150 × 250 feet. Neither the true vault nor the spherical dome was applicable to the situation because of the oval shape and the need for high acoustical standards. Grow's solution was to use a segmental vault for the middle third of the building and two half-domes with approximately spherical surfaces to cover the rounded ends. The roof of tongue-and-groove planking was originally covered with shingles, but these were recently replaced by aluminum sheathing. The central vaulted portion of the roof is supported on nine lattice trusses in the form of segmental arches with a span of 132 feet; each domed end is carried on 13 quarter-circular lattice trusses of half the main span (Fig. 13). The trusses, which are 9 feet deep, consist of paired top and bottom ribs and two paired intermediate ribs that hold the lattice web members between the paired timbers. The roof planking is nailed to purlins spanning between the trusses. The joints between ribs and web members are secured with wooden dowels the projecting ends of which were split and made tight by means of little wedges driven into the openings. The separate segments of the ribs were joined by hand-wrought iron bolts. Natural splits in the timbers were bound in green rawhide, which pulled tight as it dried and contracted. The fact that the grain of the rib timbers follows the curve of the rib indicates that they were bent to conform to the arch curve before assembly in the structure.

The roof trusses spring from a three-layered grillage of timbers set on the tops of the sandstone piers that extend around the periphery of the building. Several curious features strike the trained eye: the top surface of the pier is horizontal rather than inclined inward, as it ought to be to sustain the outward thrust of the arch, and the pier itself seems too small in its cross-sectional dimensions to take the thrust without overturning. The presence of diagonal cracks in the lower portion of the piers indicates that they are not entirely adequate to sustain the thrust, but this minor adjustment of the masonry seems insufficient for the possible structural action. The roof planking itself may have been relied on to function as a membrane, translating some of the arch thrust into tension in the wood planking. At the same time, the overturning force on each pier is reduced below the danger point because the total thrust is distributed among 44 piers. The whole structural system is radically indeterminate (see p. 58 for a definition of this term), and the stress distribution would be difficult to calculate even with the advanced mathematical and scientific techniques now available.

The bracing of the roof structure against lateral wind loads is as impressive as the arch construction. There are three separate systems:

1. Late medieval timber frame for a two-story house. Framing systems such as this constituted the basis for all timber construction in the English colonies up to the mid-eighteenth century.

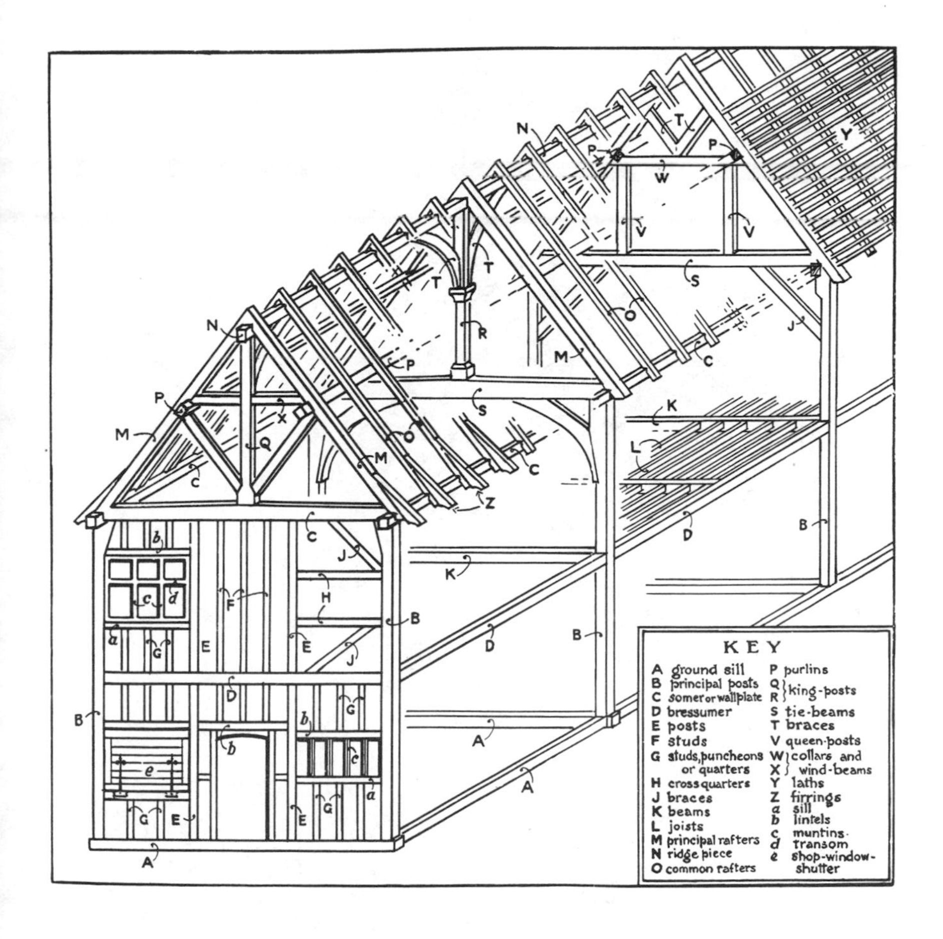

2. A New England timber frame. Restoration of the frame of the Thomas Clemence house, Manton, R.I., *ca.* 1680.

3. Old Ship Church, Hingham, Mass., 1681. Cross section of the nave showing the roof truss. Perhaps the earliest attempt in the colonies to carry a roof without intermediate supports over the wide span of a church nave.

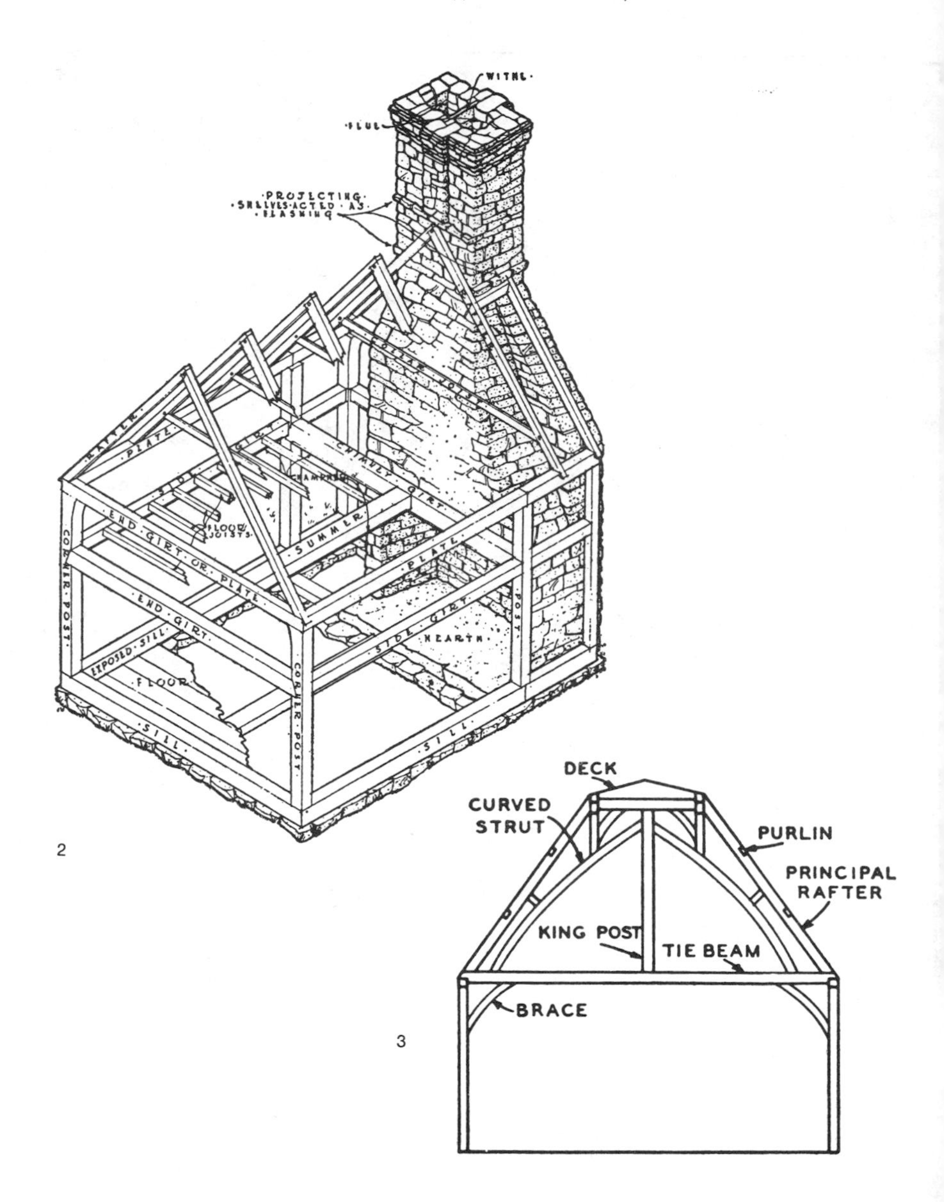

4. Independence Hall, Philadelphia, Pa., *ca.* 1732–48. Plan of the ceiling frame above the Assembly Room. Renaissance precedents lay behind this ingenious system of girders designed to support a floor over a wide enclosure.

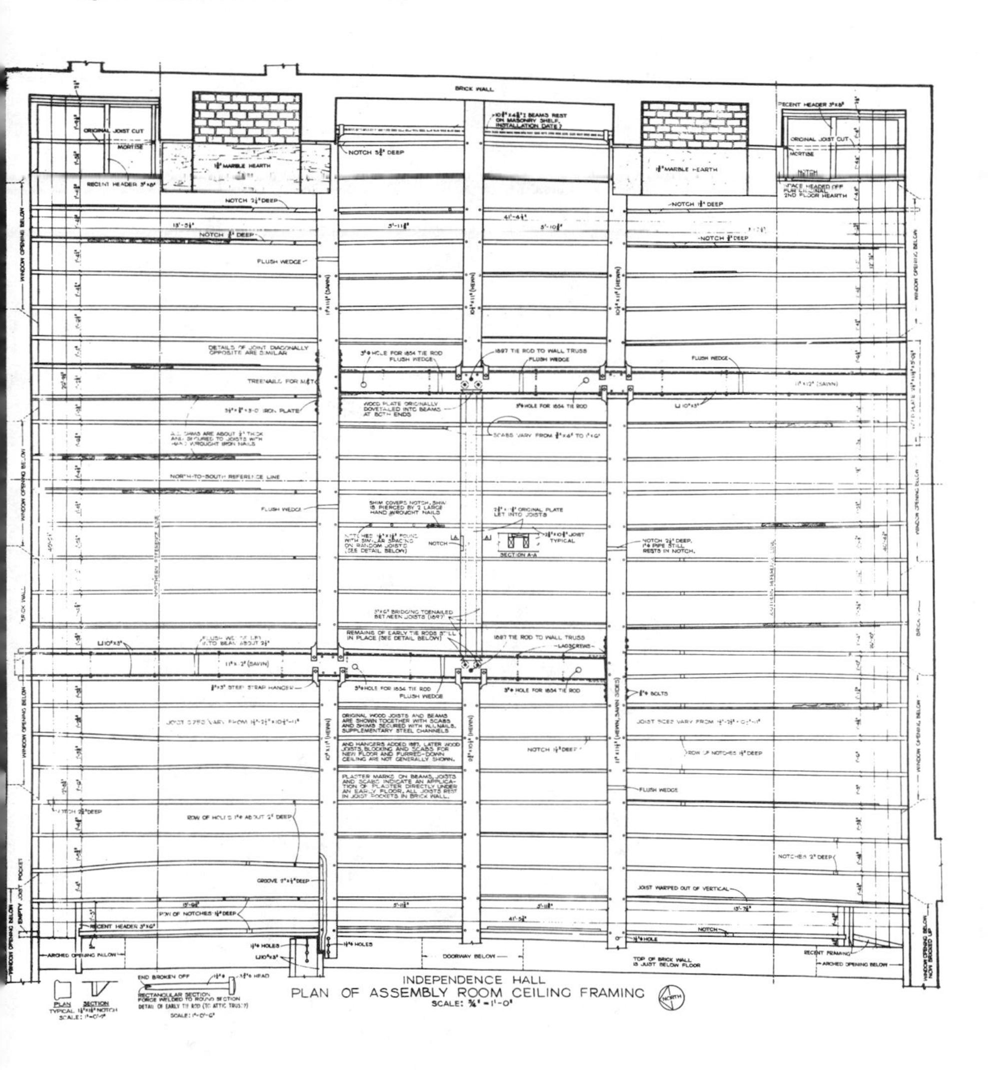

5. Independence Hall. Typical roof truss. The basic form of the braced queen-post truss has reached maturity in this structure.

6. Ursuline Convent, New Orleans, La., 1732–34. Ignace François Broutin, architect. Cross section showing the wall framing. A braced timber frame that was typical of French structural techniques in Louisiana.

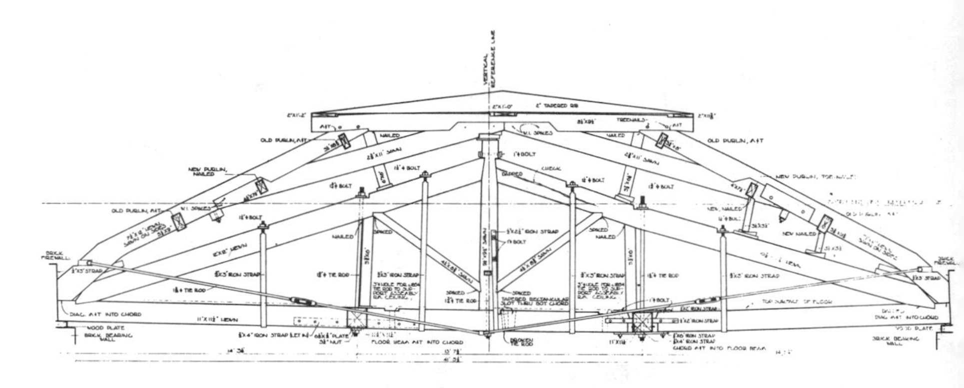

5

6

7. Bridge over the Connecticut River, Bellows Falls, Vt., 1785. Enoch Hale, builder. The first American bridge to embody a true framed construction.
8. Common brick-work bonds.

7

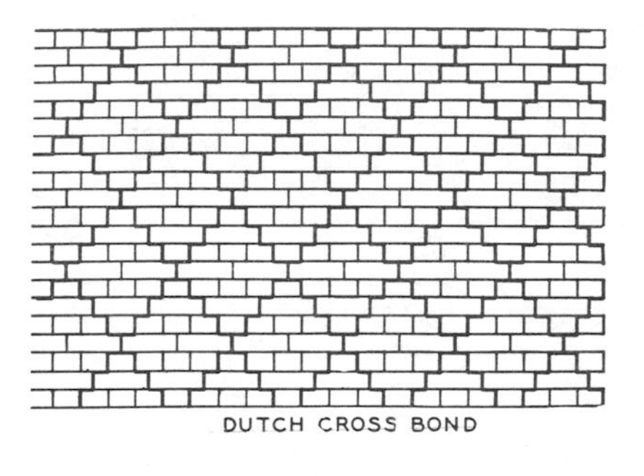

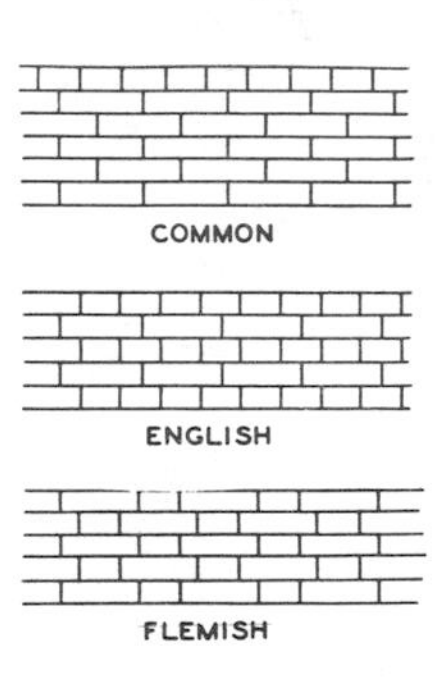

8

9. Cabildo, New Orleans, La., 1794–97. Transverse section (left) and half elevation of one of the roof trusses (right). A good example of highly developed timber framing in a former Spanish colony.

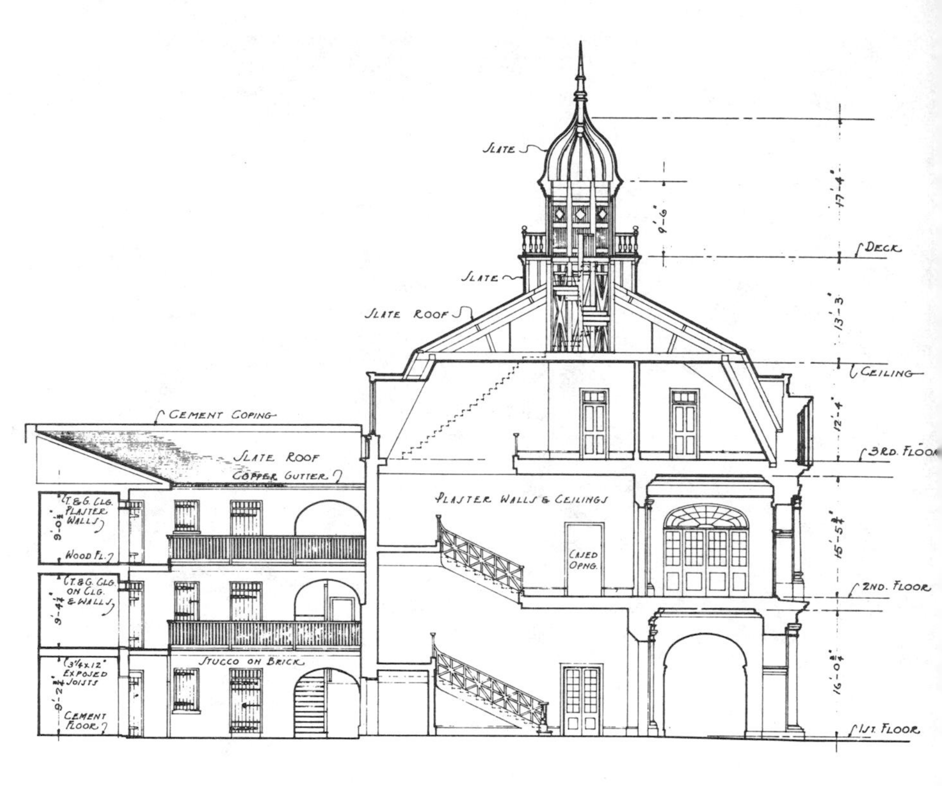

9a

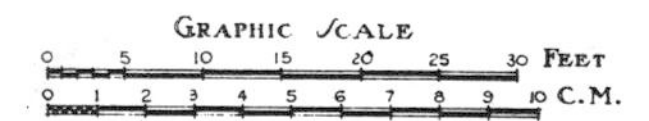

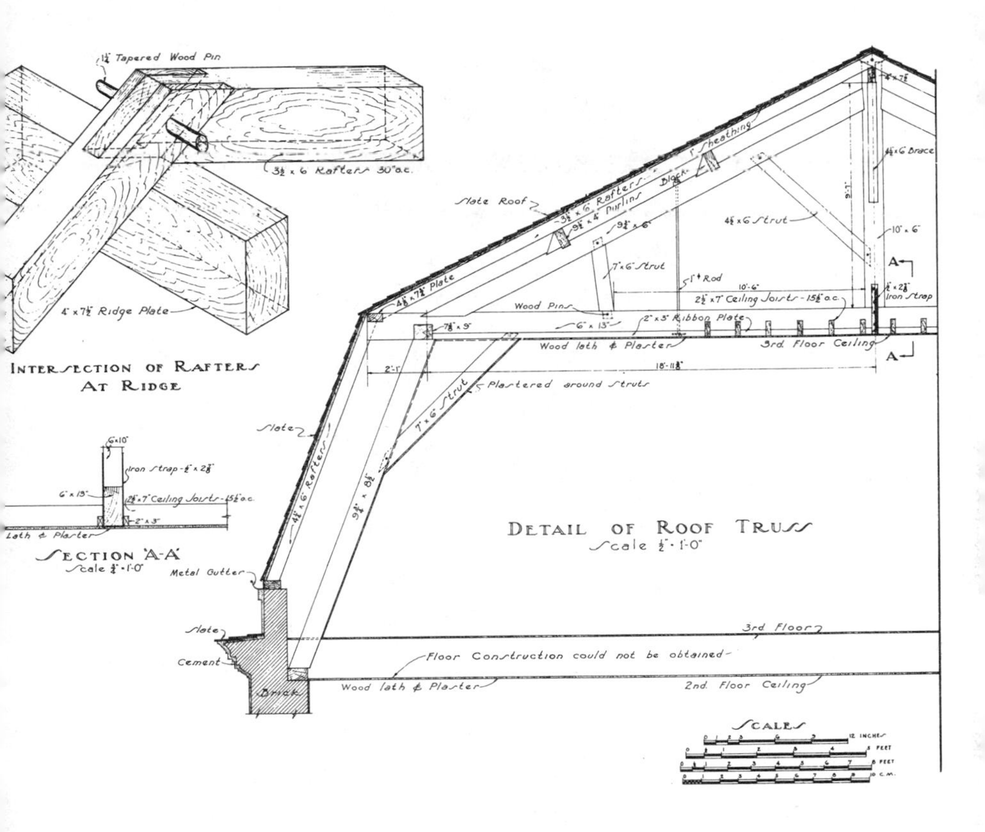
1½" Tapered Wood Pin
3½ x 6 Rafters 30" o.c.
4" x 7½" Ridge Plate
Intersection of Rafters At Ridge
Section 'A-A'
Scale ¾" = 1'-0"
Slate Roof
3½" x 6" Rafters
1" Sheathing
9¼" x 4" Purlins
Block
9¼" x 6"
4⅝ x 6 Strut
4⅝ x 6 Brace
7" x 6" Strut
1" Rod
10'-6"
2½" x 7" Ceiling Joists - 15½" o.c.
Iron Strap
Wood Pins
6" x 13"
2" x 3" Ribbon Plate
Wood lath & Plaster
3rd. Floor Ceiling
Plastered around struts
Slate
4½" x 6" Rafters
7" x 6" Strut
Detail of Roof Truss
Scale ½" = 1'-0"
Metal Gutter
Cement
Brick
3rd Floor
Floor Construction could not be obtained
2nd. Floor Ceiling
Scales

10. Balloon frame for a two-story house. The light framework could be easily erected by one man working with hammer, nails, and saw.

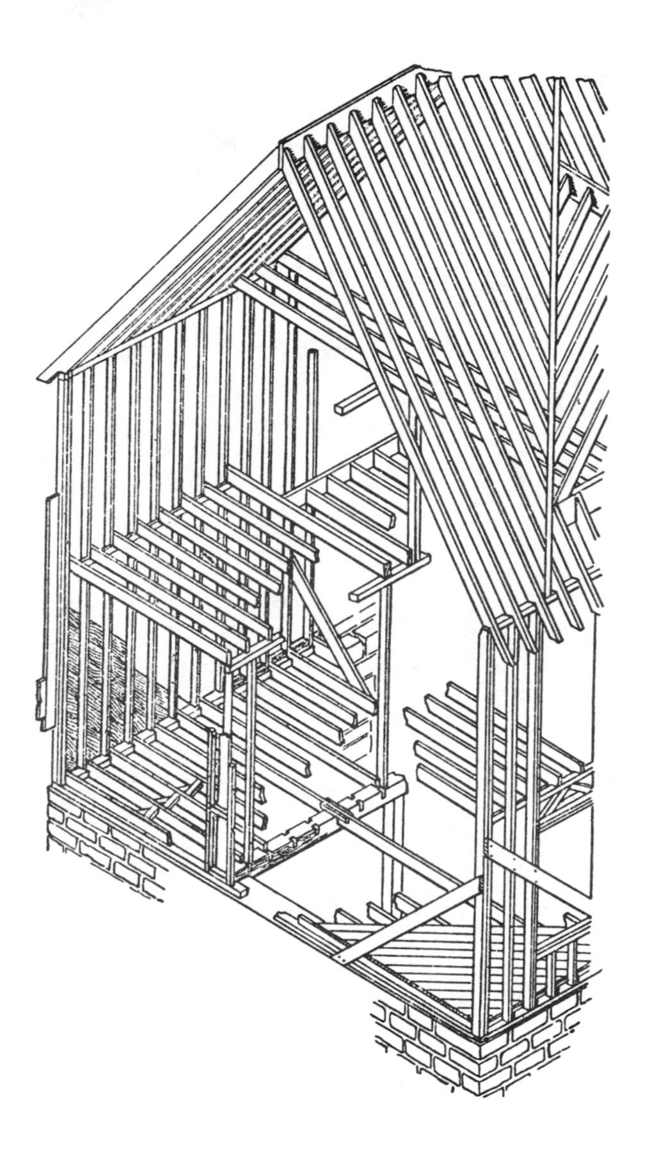

11. Inbound Freight House, Illinois Central Railroad, Chicago, Ill., 1855, 1872. Cross section showing one of the roof trusses. The composite timber and iron construction of the trusses marked the transition from wood to all-iron forms.

12. Station of the Central Vermont Railway, St. Albans, Vt., 1867–68. Interior of the train shed. This rather small shed was typical of the vaulted forms carried on arched timber trusses with iron tie rods.

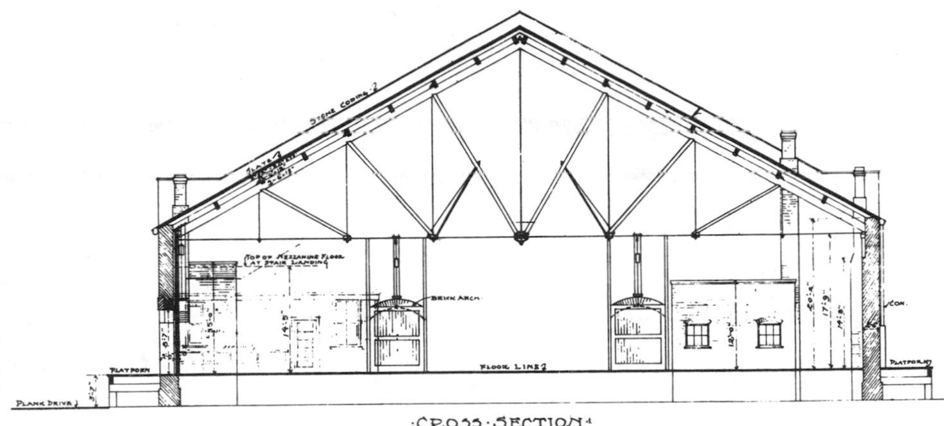

11

12

13. Mormon Tabernacle, Salt Lake City, Utah, 1863–67. Transverse section showing a roof truss. The greatest work of timber roof framing still surviving and the only one in which arched lattice trusses were used as primary supports.

14. Permanent Bridge, Schuylkill River, Philadelphia, Pa., 1798–1806. Timothy Palmer, builder. Typical of the pioneer phase of timber bridge construction, the arched ribs of the Permanent Bridge carried most of the load and the truss functioned mainly as a stiffening element.

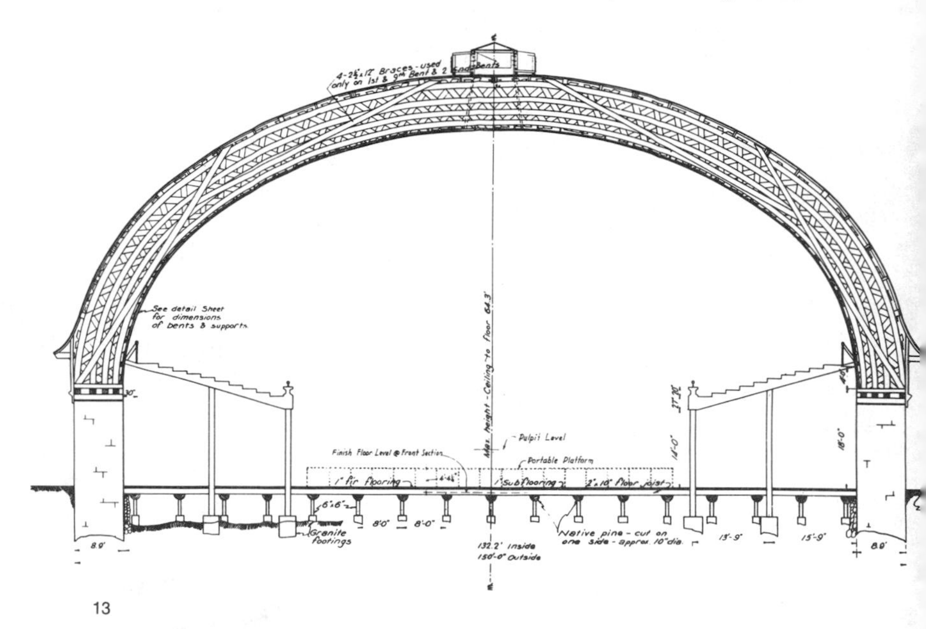

13

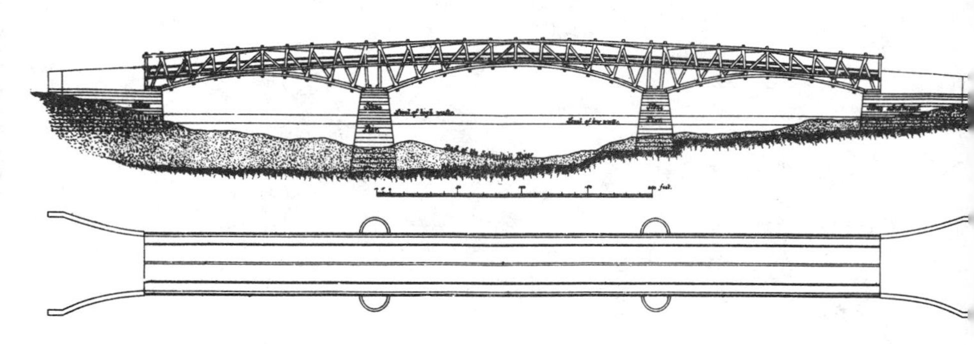

14

15. Lattice-truss bridge, Pemigewasset River, Woodstock, N.H., 1878. Town's lattice truss was the first to act entirely as a truss, but like many other forms, it was frequently combined with the arch for additional strength.

16. Truss patented by Stephen H. Long, 1830. Long's truss was the most efficient for its time, but not even Long himself fully appreciated its virtues.

15

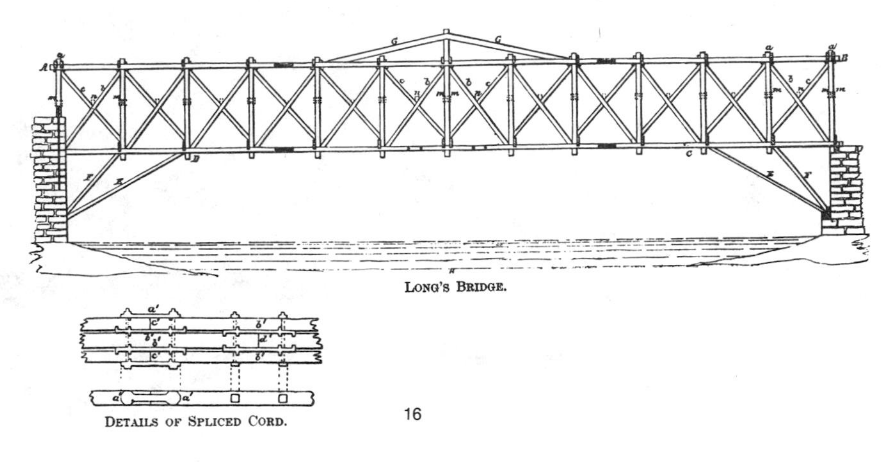

16

17. Timber bridge of the Buffalo and New York City Railroad, Genesee River, Portage, N.Y., 1851–52. Silas Seymour, engineer. Howe truss with double diagonals. The original Portage bridge was a spectacular example of timber construction with Howe trusses.

18. Highway bridge, Raystown Branch, Juniata River, Bedford Co., Pa., 1902. Howe truss with single diagonals and arch reinforcement.

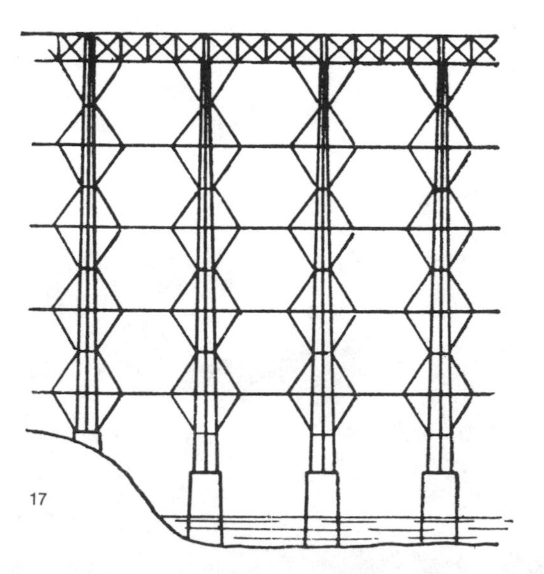
17

19. United States Sub-Treasury Building, 29 Wall Street, New York City, 1834–42. Town and Davis, Samuel Thomson, and John Frazee, architects. Cross section and plan. The expertly restored building is a masterpiece of domed and vaulted construction in masonry.

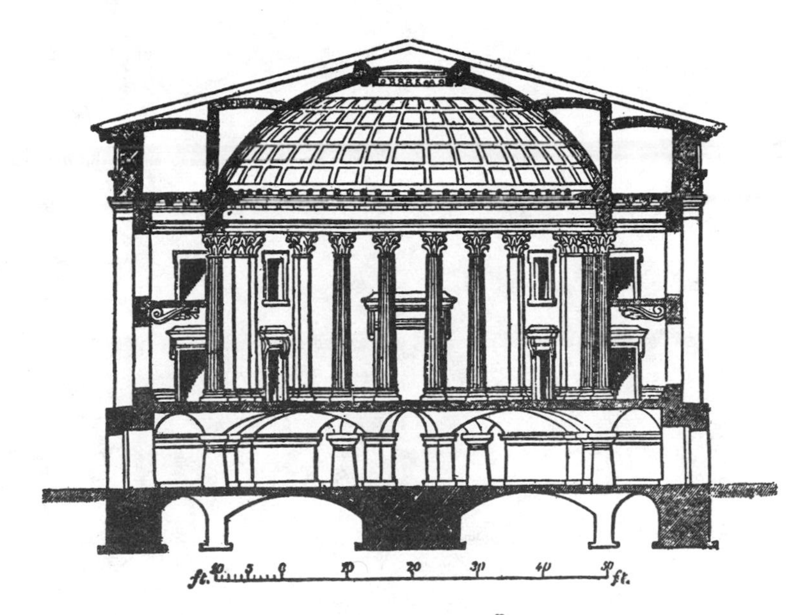

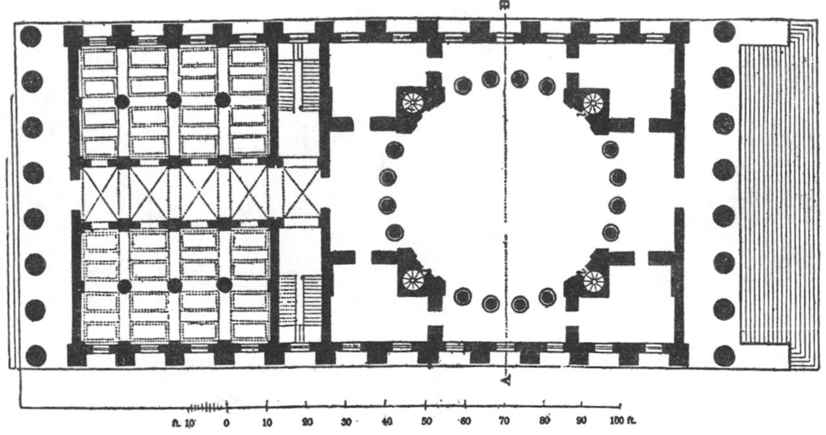

20. Founder's Hall, Girard College, Philadelphia, Pa., 1833–47. Thomas U. Walter, architect. Longitudinal section. Founder's Hall is another excellent example of the constructive genius of the leading Greek Revival architects.

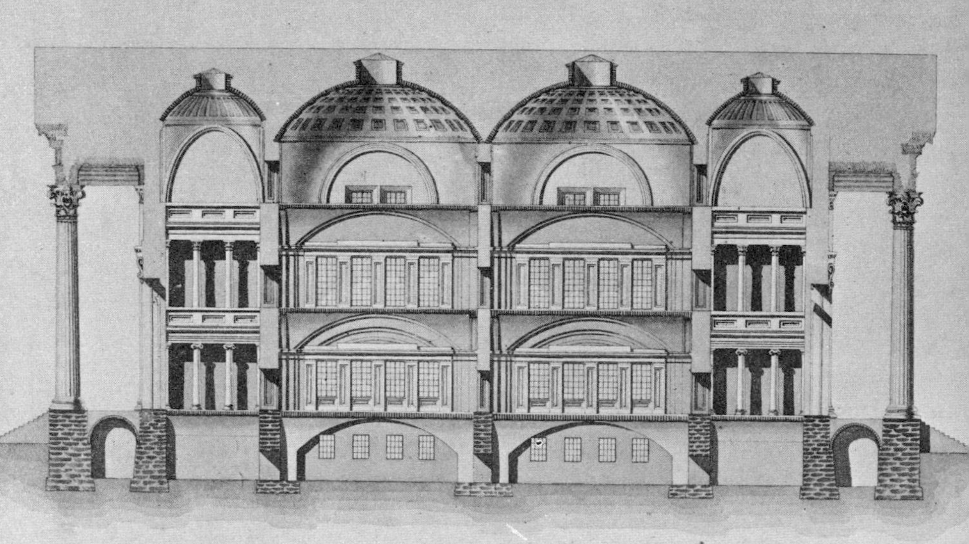

21. Cathedral of St. Louis of France, St. Louis, Mo., 1831–34. George Morton and Joseph Laveille, architects. Transverse section. One of the few examples of the Greek Revival west of the Mississippi River; the structural system of the cathedral is more noteworthy for its timber roof trusses than its masonry exterior.

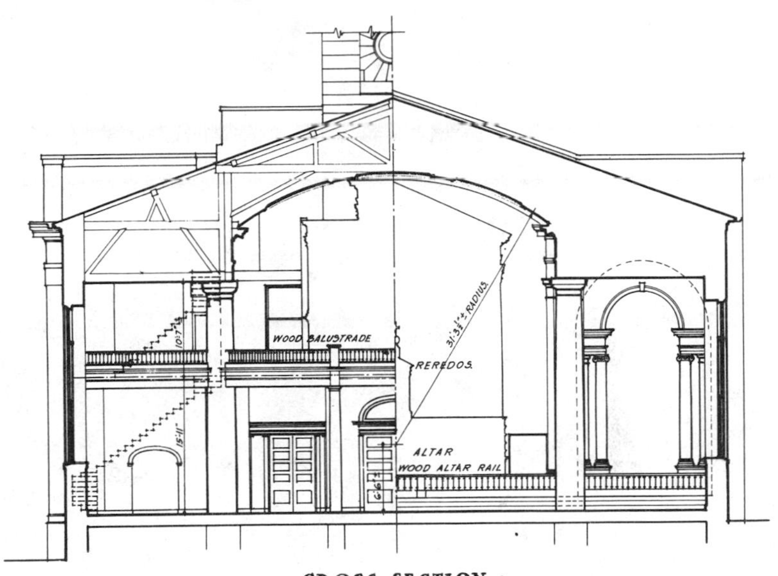

22. Aqueduct (High) Bridge, Harlem River, New York City, 1839–42. John B. Jervis, engineer. In this famous New York landmark we can clearly see the Roman precedent for most big masonry bridges in the United States.

double-diagonal bracing between trusses in planes at right angles to the arch webs; lateral struts set between the bottom ribs of the arched trusses; double-diagonal bracing between these struts, the bracing lying in planes parallel to the roof tangent planes. In addition to these primary systems, there is supplementary bracing and bridging at joints and sharp angles between members, a subsidiary system which was put together from odds and ends of scrap lumber. The ceiling of plaster bound with imbedded cattle hair is spread over lath nailed to rafters that are in turn fixed to little wooden hangers suspended below the struts, the tangential bracing, and the bottom ribs of the trusses.

All the wood in the roof and floor structure is local yellow pine. Nothing was wasted; every scrap found a useful role in the construction. Built in a desert area that at the time was fifteen hundred miles from the nearest railroad, with no manufactured materials available, every part of the tabernacle had to be made locally by hand. Grow depended entirely on his knowledge of bridge construction to guide him in the design of the roof structure: the arches, bracing, piers, and sheathing were all derived from antecedents found in timber-truss bridges. Coming shortly before iron eclipsed wood as a primary material for large structures, the tabernacle is one of the few masterpieces of building technology that remain from the heroic age of timber construction.

Chapter Four | Timber-Truss Bridges

The pile-and-beam bridge of the colonies was severely limited in length of individual span and in load capacity. The use of piling driven to position by a hand-lifted weight made it impossible to build over deep water or to found supporting bents in beds of heavy gravel, hardpan, or rock. The expansion of the turnpike network in the early years of the republic had required only that bridges be of a strength and size sufficient to carry wagon loads over navigable waterways. The coming of the railroad in 1830, however, brought with it radical changes in the criteria of bridge design: strength had to be greatly increased to sustain the weight of moving locomotives on relatively long spans; at the same time rigidity had to be increased not only against wind loads but also against the lateral sway of trains. Structural solutions adequate for mills and short-span bridges were useless in supporting moving loads over spans that frequently exceeded 100 feet in length. Truss framing offered the only solution, but it had to be developed far beyond the king-post and queen-post forms that carried the roofs of mills and churches. Until about 1840 the bridge truss was always a timber structure, a limitation that powerfully stimulated the inventive faculty of railroad bridge builders in the early years of rail expansion. Up to a point, the problem of increasing loads could be solved by enlarging timber forms, but the larger forms were vulnerable to the devastations of fire and flood, old hazards that acquired new terror when trainloads of passengers were involved.

The same factors that lay behind the use of wood for buildings operated in the case of timber bridges. Early bridges were often built crudely and with a lavish use of lumber, for they were usually erected in the expectation of frequent strengthening, enlargement, or eventual replacement. Although the masonry viaduct was dominant in England and western Europe, where the intention was to build lasting structures, the timber form was nearly universal in the eastern United States until 1865 and in the West until the early years of the new century.

The truss is an assemblage of relatively small members arranged and connected in such a way that they behave as a rigid unit. Until the end of the nineteenth century the various members were always arranged as a series of triangles because the triangle is the only figure bounded by straight lines that is rigid in itself, that is, a figure that cannot be deformed without distorting one or more of its sides. Under a moving load a truss behaves in a complex way that cannot be described in general terms applicable to every situation. When a truss supported at its ends is subject to a load, it is deflected downward, like a simple beam. The result ordinarily is that the top chord is placed under compression, the bottom chord under tension, the vertical members or posts under compression, and the diagonals under tension. But many early trusses were designed in such a way that the verticals were in tension and the diagonals in compression. Moreover, it is often impossible to predict how the diagonals will be stressed, for any one may be stressed alternately in tension and compression as the load moves over the structure. The problem was greatly complicated by the fact that the pioneer forms of the timber truss were combinations of arches and trusses in which it was impossible to determine how the load was divided between the two forms. Finally, we must remember that up to the mid-nineteenth century there were no reliable methods of analyzing the stress distribution in a truss. The development of this scientific tool in the first half of the century was largely the achievement of French mathematicians, whose work lay far beyond the educational level of the American carpenter-builder. His approach was pragmatic and crudely empirical, based on experience and an intuitive sense of the statical behavior of truss frames. Within these limits he did remarkably well; indeed, he was very largely the creator of the practical timber truss.

Properly designed trusses suitable for bridges were the invention of the sixteenth century Italian architect Andrea Palladio, but he proved to be two hundred years ahead of his time. No one attempted to put his inventions to practical use until Timothy Palmer of Newburyport, Massachusetts, adapted the arched Palladian truss to bridge construction. His first major work was the bridge built in 1792 over the Merrimack River between the towns of Salisbury and Newbury, Massachusetts. For most of its length the bridge was a traditional work of braced-beam construction on what seem to be piers of rock-filled timber cribs, although this feature is not clear in surviving drawings. The two channel spans, respectively 113 feet and 160 feet in length, were Palladian arched trusses. The deck, bowed upward to match the curve of the flattened arch, rested on a pair of arch ribs spanning between the piers. On either side of the deck there was a truss consisting of posts framed into the ribs on radial lines, with a single diagonal between pairs of adjacent posts. The load was probably sustained largely by the arch ribs, the trusses functioning mainly as stiffening elements. Diagonal bracing under the deck between the ribs provided rigidity against wind loads. The Salisbury-Newbury bridge was a pioneer work of crucial importance, for it launched a great structural development absolutely essential to modern transportation. Simple and durable in construction, a good part of it survived until 1909.

Palmer's largest and most famous work was the Permanent Bridge, built between 1798 and 1806 to carry Market Street over the Schuylkill River at Philadelphia (Fig. 14). A three-span structure, it was the first bridge composed of more than one truss in a contiguous series and possibly the first in which the structural members were covered with wood sheathing as a protection against the weather. This covering was originally known as weatherboarding. The practice was enough of a novelty for Palmer to have to defend it before the directors of the bridge company against those who argued for the less costly exposure of the structure. The deck of the bridge formed a continuous curve with a length of 550 feet on the horizontal line between abutments. It was supported by three separate arched trusses, the longest 194 feet 10 inches in span, joined by a continuous top chord which followed the curve of the deck. The posts were set on radial lines, with a single diagonal in each panel, and the piers were coursed masonry laid in mortar joints inside a timber cofferdam. The west pier established a record for American bridges of the time in that it was carried to a depth of 41 feet 9 inches below mean high water. The cofferdam for this pier was a kind originally developed by William Weston in England. The covering of the superstructure frame was designed by the Philadelphia architect Owen Biddle. The Permanent Bridge was drastically altered and enlarged in 1850 to carry the tracks of the Columbia and Pennsylvania Railroad, but the new structure lasted only 25 years before it was destroyed by fire in 1875.

At the time the Philadelphia span was nearing completion, the second of the great pioneers of timber bridge construction was beginning his career. Shortly after the turn of the century, Theodore Burr invented a combination arch and truss bridge, although he did not patent his invention until 1817. The first major bridge to embody the Burr system was built between 1803 and 1804 over the Hudson River at Waterford, New York, where it stood until fire cut it down in 1909. The four spans of the bridge varied in length from 154 to 180 feet. Each consisted of three parallel trusses, one at the center of the roadway and two at the sides, with parallel chords and two diagonals between each pair of posts. Outside of each side truss an arch of curved timbers, bolted to the posts, extended the full span between the piers. The total load of any one span was thus divided between the arch and the truss, but in such a way as to make it impossible to determine exactly what the division might have been. A practical guide for arriving at a reasonable estimate—one that Burr himself probably used—was to regard the dead load of the structure as carried by the arch and the live or traffic load by the truss.

Most of Burr's bridges were massive and redundant, combining as they did two perfectly sound structural forms. In one of his largest bridges, however, he relied on the arch alone. The five-span structure built to cross the Delaware River at Trenton, New Jersey (1804–6), was a pure arch bridge with the deck carried on transverse beams that were hung by means of wrought-iron suspenders below the underside of the arch. The main span of this original work of bridge design was 203 feet long. One of its unique features was the presence of supplementary arches, which were splayed out from the vertical plane of the main arches to provide additional wind bracing, under the deck frame at the abutments. The primary bracing against lateral sway, as in all timber bridges from Palmer's on, were the diagonals set between the stringers and transverse beams of the floor frame. The Trenton bridge was repeatedly strengthened to carry increasing rail loads until it was replaced by an iron truss in 1875. The largest of Burr's arch-and-truss bridges spanned the Susquehanna River at

McCall's Ferry, near Lancaster, Pennsylvania. The main span had the extraordinary length of 360 feet. Built from the river ice between 1814 and 1815, with reckless disregard of the most elementary safety precautions, it was destroyed by ice two years after it was completed.

Variations on the Palladian arched truss of Palmer's bridges characterized the structures of Lewis Wernwag, the first of many American builders and engineers of German birth. He began his twenty-seven-year career as a bridge builder in 1809 and secured his fame by one of his first works, The Colossus, constructed between 1809 and 1812 as the second of the Schuylkill bridges at Philadelphia. The Colossus was a single-span arched truss with a length between abutments of 340 feet and a rise of 20 feet. Like Palmer's trusses, those built by Wernwag had posts set on radial lines, but they differed from the earlier form chiefly in the presence of two diagonals in each panel instead of one and a number of wrought-iron struts acting as supplementary tension members in the panels. In spite of this elaborate truss construction, it is very likely that the major portion of both dead and traffic loads was sustained by the massive arch forms. The deck of The Colossus rested on five parallel laminated ribs with a depth of 3 feet 6 inches—enough timber in the arches to reduce the action of the trusses to that of stiffening elements. The covering of the bridge was designed by the famous Greek Revival architect Robert Mills. The durable construction of The Colossus, however, could not save it from the mortal enemy of the timber bridge, and fire consumed it in 1838, although it could have carried vehicular traffic for at least a century.

The structural excellence of Wernwag's bridges was best revealed by the span erected over the Kentucky River at Camp Nelson, Kentucky, in 1838. Three parallel arched trusses of 240-foot span successfully carried the traffic of the double roadway for ninety years. In 1927 a systematic stress analysis that was carried out to determine whether the bridge could sustain a heavier automobile traffic revealed the sound intuition with which this example of prescientific bridge construction was built. The analysis showed an exact distribution of stresses among the various members, in spite of the redundancy of the arched form: the upper chord of the truss worked in tension, the lower chord (the primary bearing member) in compression; the posts worked in tension, and the diagonals in compression. Moreover, the stress lay well within the allowable strength of the material, the maximum compressive stress being 618 pounds per square inch. On the basis of this analysis, the bridge was strengthened in 1928 and used for a number of years to carry a traffic load far beyond anything Wernwag could have anticipated. This increase in load after nearly a century of use was possible chiefly because the dead weight of the structure was much greater than the maximum live load that could be placed upon it.

The progressive improvement of timber-truss forms continued throughout the first half of the nineteenth century. Outside the mainstream of development, there was an excursion into another structural form, the cantilever, which proved to be abortive at the time, although its potentialities were later to be exploited in iron and steel. Thomas Pope, a carpenter-builder of New York, wrote the first American work on bridge construction, *A Treatise on Bridge Architecture,* which was published in 1811. Much of the book is a review of bridge types and modes of construction, with extensive data on the physical properties of various building materials. Included in the text is Pope's proposal for the first cantilever bridge design in the United States and the only one of solid girder construction in wood. Although he referred to it as a "flying lever pendant bridge," the form was that of two braced cantilevers, fixed at the abutments and composed of a series of beams resting one on top of the other, each a little longer than the one below it. This is actually a primitive form known as corbeled construction, which has been used in China for many centuries. As a matter of fact, Pope said that he derived the idea from a bridge built in 1660 at Wandipore, India. Except for the crude structures of logs on minor rural roads, the cantilever bridge did not appear in the United States until well after the advent of the iron-truss bridge (see pp. 144–48).

The great majority of timber bridges represented variations on the trusses developed by Palmer, Burr, and Wernwag until 1820, when a new invention ushered in a period of rapid progress toward more mature forms. The first truss to be free of arch action and to exert only a vertical load on its supports was patented in 1820 by Ithiel Town, a leading Greek Revival architect of New York and New England. Town's invention, known as the lattice truss, had horizontal top and bottom chords and vertical end posts, but its unique feature was the absence of intermediate posts, the web being made up of a dense array of intersecting diagonals in a tight lattice pattern (Fig. 15). This truss form quickly became popular since it offered certain immediate advantages to the bridge builder, but its active life as a timber structure was short and it began to be superseded by other forms in 1840.

The lattice truss possessed a number of virtues not shared by other types, but it also exhibited serious defects in action. Its great value was that it functioned entirely as a truss, which simplified the construction of bearings and abutments as well as that of the bridge structure itself. The separate members of the web and chords were relatively light and thus readily assembled and spiked or bolted into place. In this respect Town's invention may be regarded as the equivalent in bridges of the balloon frame in buildings. The chords of the lattice truss consisted of two closely spaced parallel lines of timbers between which the diagonal members were securely fixed. This feature, together with the practice of connecting the diagonals to each other at their intersections, made the truss tight and rigid in the vertical plane and thus able to bear fairly heavy direct loads. But the very virtues of the truss were also the cause of its defects. The thin web members and the absence of posts made the truss excessively flexible and hence unstable under lateral forces. Moreover, the inadequate rigidity in the horizontal plane gave the truss a tendency to buckle under direct loads and the vibration of traffic. Finally, like its forerunners, it was a statically indeterminate structure, that is, one embodying more unknown forces than algebraic equations that can be written to calculate them. In its wooden form the lattice truss soon lost its utility for the railroad, although it later proved to be rigid enough in iron and steel. Town proposed its construction in iron in 1831, but it was 1859 before the adaptation was actually made.

The majority of the lattice-truss bridges in wood date from the two decades between 1820 and 1840. Like the Burr-truss bridges, they were sometimes constructed with supplementary arches, in which form they continued to be built throughout the century, a few of them surviving to this day (for example, the bridge shown in Fig. 15). The longest bridge of lattice trusses was completed over the James River at Richmond, Virginia, in 1838. The over-all length of the nineteen spans was 2,820 feet, and the maximum length of an individual span 152 feet 6 inches. This great structure was destroyed by Confederate troops during the evacuation of Richmond in 1865. The construction of the lattice truss in iron and steel brought about a renewal of its active life that continued into the second decade of the twentieth century.

The rapid progress of design is forcefully demonstrated by the truss patented in 1830 by Stephen H. Long, who was at the time a colonel in the Engineer Corps of the United States Army (Fig. 16). The original design included an auxiliary king-post truss above the top chord at the four center panels (the region of maximum bending), but this redundant feature was dropped in the patents of 1836 and 1839. In the Long truss the relatively short posts were compression members and the diagonals tension, but one or the other diagonal in each panel could function in compression if the stress distribution required it. The sway bracing in the bottom frame provided another indication of Long's inventive talents: the diagonal members between each pair of transverse beams were arranged in the form of the letter K, which led to its later designation as the K-truss. Its chief advantage is that the short diagonals offer greater resistance to buckling than the long members that cross the full panel. The K-truss is now widely used as wind bracing, especially in the lateral frames of steel arch bridges, and has recently appeared as bracing in the steel frames of buildings with long bay spans.

The exact form of the panels in the Long truss, the distribution of stress, and the careful proportioning of individual members reveal that Long's understanding of truss design was well in advance of that of his predecessors. This accurate knowledge most impressed the great German engineer and theorist Karl Culmann on his visit to the United States in 1849. The questions that immediately attracted

Culmann's attention still perplex the historian: What were the origins of Long's ideas, and how much did he know of the new European theory of truss design? We can do little more than describe the leading pioneer treatises on the subject and suggest a possible background to Long's work.

The primary source of practical bridge design in Europe and of subsequent theoretical works on both sides of the Atlantic was Emiland Gauthey's *Traité de la Construction des Ponts,* edited and published posthumously, 1809–13, by his nephew Louis Marie Navier. Gauthey's theory of truss design rested on two principles, called by him equilibrium of position and equilibrium of resistance. The first represents a pioneer attempt to resolve the loads on a bridge truss into the component forces acting in the individual truss members. The second principle refers to the elastic properties of wood and to its tensile and compressive strength, and thus has to do with the size and proportions of the individual member. Navier was one of the brilliant mathematicians at the École Polytechnique in Paris and the foremost mathematical theorist of structural science. His technical learning and his rigorous, highly abstract thought were far beyond the mathematical competence of American builders in the first half of the nineteenth century. The ideas of the French theorists would have remained unavailable were it not for the appearance in English of a number of simplified epitomes that came in rapid succession after 1820. The first of these was Thomas Tredgold's *Elementary Principles of Carpentry,* originally published in England in 1820 and in the United States in 1837. Chief among its American successors were Jacob Bigelow's *Elements of Technology* (1831) and James Renwick's *Elements of Mechanics* (1832). These works were undoubtedly known to Long by the date of his 1836 patent, as they were to his foremost inheritors in truss design, but the question of the source of Long's 1830 truss remains obscure. The leading builder of timber-truss bridges in Europe at the time was Carl Friedrich Wiebeking of Bavaria, but it is doubtful whether his *Beyträge zur Brückenbaukunde* was known outside the region of the German dialects.

The first bridge with Long trusses, known as Jackson Bridge, was built in 1829 for the Baltimore and Ohio Railroad probably at Baltimore. The first with K-truss bracing was constructed by the United States Army Engineers in 1830 to span the Mattawamkeag River in Maine. After that date the record of Long-truss bridges is extremely thin. In spite of their advanced form, they were less popular than those of any other inventor. But this did not prevent Long from further experiments. He was granted two more patents in 1847 and 1858, the latter for an inefficient, extremely redundant arch-reinforced truss that stood far below the level of his first design. The existence of this curiosity at the end of Long's career indicates that the truss was still inadequately understood at the mid-century and that the 1830 truss may have rested less on science than on ingenious guesswork.

The inventor of the most widely used bridge truss in the nineteenth century was William Howe of Massachusetts. He built his first bridge for the Boston and Albany Railroad in 1838 and received two patents on the structural design in 1840. The truss was a curious form with parallel chords, triple-intersection diagonals of wood, and vertical tension members of wrought iron. The design suggests a compromise between the multiple diagonals of the lattice truss and the elegant simplicity of Long's invention. In collaboration with the railroad builder Amasa Stone, Howe developed what came to be the standard form, which he patented in 1841: the parallel chords and iron rods remained the same, but the diagonals were reduced to two in each panel (Fig. 17). It was this form that was everywhere known as the Howe truss and that enjoyed the greatest popularity throughout the subsequent history of the timber bridge. In appearance it was like the Long truss and the Burr truss without the arch, but it differed from both in that the verticals were always tension members. Many Howe trusses, however, were built entirely of wood and were thus not perfectly distinguishable from the Long truss. In the East they were usually built with wrought-iron verticals and cast-iron bearing blocks; in the West, where iron was costly, they were of timber throughout.

The Howe truss was rapidly adopted for railroad bridges of great size. The most spectacular was built between 1851 and 1852 to carry the tracks of the Buffalo and New York City Railroad over the deep gorge of the Genesee River at Portage, New York (Fig. 17). The structure was designed by Silas Seymour, chief engineer of the railroad. Although the individual spans were modest in size, with a length of about 50 feet, the rail stood at a height of 234 feet above the river bed. The bridge was destroyed by fire in 1875 and replaced by an iron structure the same year (see p. 97).

By the time the Portage bridge was completed, builders of Howe-truss spans had introduced two variations in the design of 1841. One was a simplification that reduced the number of diagonals to one in each panel, with the result that the truss became a determinate structure and engineers could make an exact analysis of forces in the various members. The second was the adoption of the common practice of adding supplementary arches to the truss structure (Fig. 18). Howe was granted a patent on this variation in 1846. The arch-reinforced double-diagonal form was used for the largest bridges. One of the earliest was the all-timber structure built between 1848 and 1849 to carry the tracks of the Pennsylvania Railroad over the Susquehanna River at Rockville, Pennsylvania. Its length of 3,800 feet probably established a record for the time. The Rockville span was replaced in 1877 by an iron structure, which was in turn replaced in 1902 by the present arch bridge of masonry and concrete. The greatest of all Howe-truss spans was the mile-long bridge built by George A. Parker between 1862 and 1866 for the Susquehanna crossing of the Philadelphia, Wilmington and Baltimore Railroad at Havre de Grace, Maryland. This bridge was also replaced by an iron span in 1877, about the time that the railroad company was absorbed by the Pennsylvania. The arch-reinforced truss with single diagonals was increasingly confined to highway spans, where it enjoyed an active life until well into the twentieth century. The Juniata River bridge in Bedford County, Pennsylvania, completed in 1902, is a late eastern example (Fig. 18). The timber bridge survived longest in the Pacific Northwest, where the Howe truss continued its dominant role. In Oregon, where wood is abundant and automobile traffic relatively light, timber bridges with Howe trusses were built as a matter of practical policy until 1952.

A special type of timber framing which is a peculiarly American technique is the trestlework designed to support rail lines over deep valleys of V-section. Trestlework is a continuous frame of inward-sloping posts tied together by longitudinal and transverse beams, the whole system braced in both directions by diagonal members. The intention was to provide a rigid supporting structure under the entire length of the bridge, not only to carry traffic loads but to withstand wind action, floods, train sway, and the horizontal forces associated with the braking and acceleration of trains and with their movement over bridges on curving alignments. Trestlework appeared in the early phase of the timber bridge but reached its most spectacular development on the western lines. The first company to build in the western mountains, from Oakland, California, to Ogden, Utah, was the Central Pacific, which was constructed between 1862 and 1869. Theodore Judah and Samuel M. Skerry, successively chief engineers of the road, were responsible for amazing feats of trestle construction over the canyon-like ravines cut into the steep slopes of the Sierra Nevada Mountains. The construction of the Cascade Division of the Great Northern Railway between 1890 and 1893 required equally daring achievements, the most impressive of which was the long curving trestle fixed between the mountain walls along Martins Creek at Tye, Washington. Trestlework of this kind has mostly been replaced by steel bents and girders, but a few examples still survive on logging railroads.

Chapter Five | Masonry Construction: Buildings and Bridges

Since iron-framed construction was derived in good part from timber prototypes, the discussion of its development logically belongs in this place. There are several historical reasons, however, why the consideration of masonry building must precede it. In the first place, construction in brick and stone is far older than any other form of durable building, having reached a highly developed state in classical antiquity. Moreover, masonry was the dominant material for a kind of building that was rapidly coming to prominence in the early decades of the republic. Timber construction belonged in good part to vernacular building—mills, warehouses, bridges, and the like—constructed for strictly utilitarian ends, whereas masonry was the obvious choice for buildings consciously designed for expressive and symbolic as well as practical purposes. Until the advent of steel framing, large ecclesiastical, governmental, and even commercial buildings required masonry construction not only for formal and traditional reasons but for structural ones as well. Iron was not introduced until the climax of large-scale masonry construction around the time of the Civil War, and then only for certain interior structural elements, especially in the support of domes and vaults, and is thus to be regarded as a supplementary material incorporated into the body of an essentially masonry building.

Although the design of large domed or vaulted structures might require a high degree of architectural talent, such ability differed markedly from the special combination of boldness and empirical knowledge demanded by the construction of a timber railroad bridge over a major waterway. The principles of masonry construction that exercised the greatest influence on the building arts were developed by Greco-Roman architects in the early centuries of the empire and first codified in mathematical form by Leone Battista Alberti in 1485. The subsequent development of high ribbed domes by Renaissance and Baroque architects gave the nineteenth-century builder an extensive body of well-tested precedent to guide him in the construction of masonry building derived from the classical tradition. His primary concern was to find methods by which traditional techniques could be adapted to the demands of commerce and government.

Masonry construction in the first half-century of the American republic was almost entirely bound up with the Greek Revival, although the appropriate techniques were continuous with English and colonial precedents. The building customarily regarded as initiating the Greek Revival in the United States and as best representing the pioneer phase of all-masonry construction was the Bank of Pennsylvania, designed by Benjamin Latrobe and built in Philadelphia between 1798 and 1801. Typical of the early work, the primary bearing members were Ionic colonnades at front and rear and solid walls of carefully dressed stone at the sides. A low dome with a circular skylight at the crown covered the central area. William Strickland's Bank of the United States, built between 1818 and 1824 also in Philadelphia, is more advanced by virtue of the Roman vault of solid brickwork that constitutes the ceiling of the main banking floor. The vertical load of this vault is sustained by six pairs of Ionic columns, and the horizontal thrust is transmitted to the massive side walls of marble by the beams of the aisle ceilings. Although thoroughly traditional in form, the structural nicety of the building reveals the union of engineering and architectural genius that always distinguished Strickland's work. The United States Bank was the Philadelphia Custom House from 1845 to 1934 and is now the headquarters of the Independence National Historical Park.

The main line of structural progress in masonry lay in the development of vaults and domes, since bearing walls and column-and-lintel systems are so basic in form that they cannot be refined beyond the Roman level. At the same time that commercial requirements and increasing wealth made this progress necessary and possible, the growing skill of the builder was available to meet the demand. The rapid evolution of vaulted forms is best demonstrated by the Sub-Treasury Building on Wall Street in New York, originally planned in 1832 and constructed between 1834 and 1842 (Fig. 19). The initial plans were prepared by Town and Davis but were later altered by Samuel Thomson, who resigned in 1835 and was replaced by the sculptor John Frazee. The result of all this effort is an American masterpiece of masonry construction, fortunately still standing in restored condition.

There were a number of local factors that made the Sub-Treasury appropriate to New York and a likely product of its milieu. The city had become first in population by 1810; its rise as a port was spectacular, and by 1825 its foreign trade equalled nearly half that of the entire nation. Domestic and foreign commerce powerfully stimulated the building of offices, banks, stores, warehouses, and factories. The demand for new construction received further impetus from the disastrous fire of December, 1835, which destroyed seven hundred buildings in the Wall Street area. Utilitarian needs and the formal principles of the Greek Revival combined to produce a simple structure of masonry walls and internal timber frame that was nicely adapted to a wide variety of commercial needs. The great wealth of the city and its important political role in the early years of the federal government established a tradition of public building that was soon to match Philadelphia's proud architectural heritage.

The Sub-Treasury Building was originally built as the New York Custom House but was converted to treasury use in 1862, a function which it retained until 1939, when it was designated as a National Historic Shrine. The building is a rectangle in plan with over-all dimensions of 90 × 178 feet. The gable roof is carried on eight Doric columns of square section at each end and solid buttressed walls along the sides. This simple Doric temple, as it appears from the streets, gives no indication of the structural intricacy of the interior. The dominant visual feature is a low marble dome of 60-foot diameter that covers the central rotunda of the main floor, but this is entirely hidden on the outside by the gable roof (Fig. 19). Behind it (toward the north) and in the basement are a number of offices and enclosures for the storage of coin and bullion, all carefully designed for permanence, fire-resistance, and maximum security. These rooms are enclosed in solid brick partitions and covered with low groin vaults of brick supported at their corners either by the wall buttresses or by columns massive enough to sustain the horizontal thrust of the vaults as well as the vertical load.

The domed rotunda of the main floor posed the primary structural problem and led to the most ingenious solutions to support its load and thrust. The dome rests on 16 marble columns arranged in groups of four between the huge corner piers, which are large enough to carry their load and to provide space within them for helical stairways. The horizontal thrust of the dome is further sustained by four piers at the second-floor level and by flattened vaults between the piers at the level of the second-floor ceiling. Curious features of the top-floor construction are the extremely flattened vaults of brick set in the space between the outer surface of the dome and the roof, apparently to increase the rigidity of the dome web and the walls supporting the roof as well as to function as ceilings for the rooms immediately under the sloping roof planes. The chief difficulty, however, was the support of the rotunda floor, a great 60-foot disc of granite slabs. At the basement-floor level under the center of this area is a massive ring pier of marble surrounded by a flattened ring vault of granite carried on its outer circumference by 16 heavy, squat marble columns set in groups of four between corner piers. The vault supports the rotunda floor, and the ring pier and the circumferential columns and piers carry both the vertical load and the radial thrust of the vault. This remarkable system of vaulting is repeated in the subbasement below grade level. The immediate American antecedent of this technique of supporting superimposed floor and column loads by circular vaults is undoubtedly the subfloor vaulting of the United States Capitol in Washington (see pp. 68–69).

The conversion of the building to treasury use in 1862 required various additions to provide adequate protection for the storage of gold and silver. The limestone piers and walls of the basement vaults (in the banking sense of the word) were lined with cast-iron plates 1¾ inches thick. The public character of the silver vault in the collector's room adjacent to the rotunda led to extraordinary precautions, possibly unique in the history of devices for frustrating robbery. The inner brick walls were lined with two thicknesses of cast-iron plates fixed to little wrought-iron I-beams laid on diagonal lines. Between the plates there is a layer of cast-iron spheres like ball bearings set into hemispherical pockets cut into the plates. The idea was to prevent an energetic thief from drilling through the plates: when the drill bit struck the spheres, it was thought that they would rotate without advancing the bit. Since no one has ever attempted to break into the room in this way, the validity of the idea has never been tested. The theory may be naïve, but it is indicative of the highly functional character of the building and of the structural ingenuity that was lavished on its design.

A comparable work of masonry vaulting in Philadelphia is Founder's Hall of Girard College, the various buildings of which were constructed between 1833 and 1847 from the designs of Thomas Ustwick Walter (Fig. 20). The Hall is larger than the Sub-Treasury, with dimensions inside the colonnades of 111 × 169 feet, but the vaulting is more traditional in character. The main structure is an enclosure of solid bearing walls set inside a peristyle of marble Corinthian columns 55 feet high. Each of the three floors is divided into entrance halls at the front and rear and four classrooms exactly square in plan. On the first two floors the rooms are covered by flattened groin vaults of brick whose horizontal thrust is taken by the heavy square-section piers at the corners of the room. At the third floor, where the height under the roof is greater than that between the lower floors, Walter used the additional space to cover the rooms with domes carried to the corner piers by pendentives. The groin vaults and domes considerably increased the open area of the interior spaces through the transmission of floor loads to piers rather than to interior bearing walls. The partitions could then be reduced to light screens without bearing function. The thrust of the domes and vaults is taken by the corner piers in the rooms, the buttresses in the external walls, and the weight of the entablatures above the colonnades. Like that of the Sub-Treasury, the formal design of Founder's Hall belongs to the Greek Revival, but its internal structure goes far beyond anything the Greeks attempted. Talbot Hamlin described it as "a building full of ingenuity and invention, daringly conceived and owing its final form chiefly to the perfect blend of structural engineering and aesthetic design."[1] Thomas Walter's long period of study in the natural sciences gave him this empirical and functional turn of mind.

The tradition of heavy masonry construction in domed and vaulted forms is summed up by the structural system of the United States

[1] Talbot Hamlin, *Greek Revival Architecture in America* (New York: Oxford University Press, 1944), p. 86.

Capitol. The greatest of our national monuments was designed, redesigned, built, destroyed, and rebuilt over a long period of time. The initial competition was held in 1792, the cornerstone laid in 1793, and the building as originally planned was substantially completed by 1814, when it was burned by British troops. Reconstruction and revision continued at intervals up to 1864, when the great dome was at last completed. The original building was mainly the work of William Thornton, Stephen Hallet, George Hadfield, and Benjamin Latrobe; the reconstruction after the fire of 1814 followed the designs of Latrobe and Charles Bulfinch; the dome was the achievement of Thomas Walter and his engineering associates, Montgomery Meigs and August Schoenborn. The floor loads of the Senate wing, built according to Latrobe's design, are carried on an unusual system of parallel transverse vaults and semidomes, particularly massive and powerful in the space under the old Senate chamber and rotunda. Most remarkable of the structural features dating from the Latrobe-Bulfinch period is the system of radial groin vaults and annular vaults forming double concentric rings supporting the central rotunda of the main floor. The idea for this intricate complex of masonry vaulting probably originated with Latrobe. The dome, on the other hand, came late enough for its builders to make use of iron construction, and we will defer its treatment to a later chapter (see pp. 90–91). The influence of the Capitol was enormous, and its Baroque forms and masonry construction were repeated with variations in federal buildings, state capitals, and county court houses throughout the nation up to the time of World War I.

Multistory structures like the Sub-Treasury and Founder's Hall require masonry vaulting to carry floor loads over interior enclosures. In single-story buildings, however, a vault may be used only as a formal element and can thus be built of plaster on lath fixed to a wood frame, as in the case of large colonial churches. Masonry construction combined with timber truss framing reached its culmination in the old Cathedral of St. Louis, Missouri, built between 1831 and 1834 from the designs of George Morton and Joseph Laveille. The renovated church is now preserved as part of the Jefferson National Expansion Memorial Park along the St. Louis waterfront. The front wall of the cathedral and a 27-foot length of the side walls, the four Doric columns at the entrance, the pediment, and the square-section tower are built of dressed limestone blocks. The remainder of the wall area, however, consists of irregular blocks laid roughly to course, a construction that seems crude beside the finished craftsmanship of the dressed masonry. The dominant feature is the belfry tower, which rises 90 feet above grade from brick footings that are independent of the rest of the structure. The thick walls of the tower support a 45-foot spire with an internal frame of eight posts joined by peripheral beams and braced by double diagonals in the panels.

The nave of the cathedral is covered by a flattened vault of plaster, the two side aisles by conventional plaster ceilings (Fig. 21). The hidden space between these plaster shells and the gable roof contains the most elaborate structural feature of the building, a well-developed system of timber trusses carried by the side walls and the radial-brick columns between the nave and the aisles. These trusses have the usual double function of supporting the roof above them and the nave and aisle ceilings below. The plaster of the nave vault is applied to hand-split lath nailed to a frame made up of parallel sets of planks laid on edge in such a way that their under edges form a polygon that approximately circumscribes the segmental curve of the vault. The planks may be thought of as primitive arch ribs that bear on the longitudinal stringers laid over the tops of the brick columns. The roof rests on walnut trusses of 48-foot span in the traditional form of the braced king-post. The plaster and lath of the aisle ceilings are carried on joists set between transverse braced beams that span from the columns to the side walls. The roof trusses are joined longitudinally by two braced frames that extend along the sides of the nave over the tops of the columns. These frames have the form of the Warren truss, patented in London as a new type in 1848, and may represent the earliest use of the form (see pp. 100–101). The chords are the double row of stringers above the tops of the columns, the vertical members are the posts carrying the ends of the roof trusses to the columns, and single diagonals slope in opposite directions in alternate panels.

Masonry Construction: Buildings and Bridges

Most large structures of masonry and timber belonged to the category of vernacular building, like the brick-walled mills which began to appear, as we noted, around 1810. The height of industrial buildings was steadily increased as the demand for floor space grew. This expansion in size, however, was not accompanied by any structural innovation but rather by the simple practice of enlarging the thickness of walls or buttresses to the point where the sacrifice of space eventually became intolerable. By the mid-century there were factories high enough to be regarded as skyscrapers had they been built in tower-like form in the crowded commercial areas of the cities. One amazing example was the steam sugar refinery of Joseph S. Levering and Company in Philadelphia, completed and in operation by 1852. Records of this structure are scanty, but the appearance of the fenestration in a surviving print suggests that the two main blocks were 10 stories high. The walls were undoubtedly brick, and if the traditional dimensioning was followed, they were about five feet thick at the base. The upward rise of masonry towers was to continue until the last decade of the century, especially in New York, but by that date the iron-framed curtain-walled skyscraper had been created, and the day of the traditional multistory masonry bearing wall was done.

The arch bridge of stone has always been a costly type, with the consequence that in the United States, where wages are high and materials abundant, the form has never been as popular as it has been in Europe. By the time rail traffic loads reached the point where timber spans could no longer carry them, iron trusses were sufficiently developed to take over the job. Eventually concrete supplanted both materials for highway bridges and short rail spans. But masonry bridges are extremely durable structures, and a surprising number of the few that were built survive in active use today. The arcaded bridge and aqueduct were among the conspicuous achievements of Roman builders, who nearly always used the semicircular or full-centered arch, which seemed the natural choice to railroad builders of the nineteenth century. The ideal arch form, however, is that of an inverted catenary, the curve of a freely falling, perfectly flexible cord suspended at its ends, or a parabola, which corresponds closely to the catenary; but it was not until reinforced concrete construction was fairly well advanced that arches could be built economically in the parabolic form. Although the engineers often revealed great skill and daring in the construction of long railroad viaducts, they were closely bound to the accumulated experience of a two-thousand-year tradition.

There is little evidence, as we have noted, for the building of masonry bridges in the colonies. By 1800 modest spans were being constructed of crudely dressed stone in New England and the Middle Atlantic states. In some cases the spandrel walls of the bridge were built of rubble masonry without mortar, only the wedge-shaped voussoirs of the arch proper being dressed to provide matching surfaces. The first work of finished design and execution, with carefully dressed stone and uniform mortar joints, is the Carrollton Viaduct, constructed by the Baltimore and Ohio Railroad at Baltimore, Maryland, in 1829. The engineers of the single semicircular arch of 100-foot span were Jonathan Knight and Caspar Wever of the company's engineering staff, and the architect was James Lloyd. The Carrollton bridge is still in use and is thus the oldest rail structure in the United States.

The railroad, which was the only transportation system that required the large masonry bridge, was also the only one that could afford to sponsor it. Within a few years of the Carrollton Viaduct, the B. and O. pioneered again when it built the first multispan masonry bridge for American railroad use and the first in the United States to be laid on a curving alignment. This masterpiece of the mason's art is the Thomas Viaduct over the Patapsco River at Relay, Maryland, opened to traffic in 1835 and named after Philip E. Thomas, the first president of the railroad. The chief engineer of the project was Benjamin Latrobe II, son of the celebrated Greek Revival architect. The Thomas Viaduct, divided into eight granite-faced semicircular arches of 60-foot span, extends 600 feet in over-all length measured along the arc, and the double-track rail line stands 65 feet above mean water level. The unique feature of the structure is the placement of the lateral pier faces, which are laid out on radial lines because of the curving plan, each pier thus having a wedge-shaped horizontal section. The bridge was built for locomotives that weighed little more than the 3½ tons of the B. and O.'s "York," but because the mass of material in the arches and piers is great enough to absorb the enormous compressive loads and bending forces exerted by modern locomotives, it continues to carry successfully the passenger and freight traffic of the railroad's eastern main line. The Thomas Viaduct is a superb work of architecture as well as of engineering: the simple moldings and the rough-faced granite blocks carefully set in uniform joints give the massive span great power and dignity.

The chief exception to railroad sponsorship among early masonry structures is Aqueduct Bridge, which crosses the Harlem River at 174th Street in New York City (Fig. 22). Popularly known as High Bridge, it was built between 1839 and 1842 under the direction of John B. Jervis to carry water from the original Croton Dam to the city's distributing reservoir at Fifth Avenue and Forty-second Street, now the site of the New York Public Library. The total length of Aqueduct Bridge is nearly 1,200 feet, originally divided into fifteen arches varying in span from 50 to 80 feet. The high palisade that forms the west (Manhattan) bank of the river fixed the over-all height at 121 feet above mean water level and gave the piers the extraordinary height that earned the bridge its popular name. After 1848 the structure carried three large water mains, but these were long ago superseded by the present underground siphons. The important structural innovation of the Aqueduct is the elaborate system of load-distributing walls within the hollow volume enclosed by the arch barrels, deck, and spandrel walls. In most bridges this space was filled with rubble as a support for the deck between the spandrel walls. In the Aqueduct span the extremely heavy water load in the deck trough was transmitted to the arch barrels by a system of longitudinal and transverse brick walls braced by monolithic diagonal struts in the horizontal plane across the corners of the intersecting walls. In 1937 the five arches spanning the river and the New York Central Railroad tracks were replaced by a single steel arch, but the rest of the bridge remains substantially as built.

For its rich texture and slender proportions, the handsomest masonry bridge dating from the pioneer phase of rail construction is the Starrucca Viaduct of the New York and Erie Railroad, built in 1848 to cross the wide valley of Starrucca Creek near Susquehanna, Pennsylvania (Fig. 23). With the lavish backing of British capital, the railroad company spared no expense to build a first-class double-track line through the Pocono and Allegheny Mountains. The $320,000 cost of the Starrucca Viaduct earned it the reputation of being the most expensive railroad bridge in the world at the time, a characteristic perhaps appropriate to what was then the longest rail line in the world, extending from Jersey City to Dunkirk, New York. The engineer in charge of the project was Julius W. Adams, and the contractor was James P. Kirkwood, who had been trained by George W. Whistler, chief engineer of the Western Railroad of Massachusetts and an expert on masonry arch bridges in mountain settings. Starrucca is 1,040 feet long between abutments and stands 100 feet above the water level of the stream. The 50-foot span of the seventeen arches is typical of the masonry viaduct and was seldom exceeded even in concrete for rail bridges. Starrucca is extremely severe in appearance because of the absence of moldings, its beauty arising from the slender proportions of the piers and the texture of its facing of local bluestone (a dense argillaceous sandstone). There are several novel elements in the structural system of the viaduct: three internal walls of brick extending continuously from end to end of the structure transmit the deck load to the arch barrels; the huge pier footings, each measuring 19 × 40 feet in plan, are concrete; and the bluestone slabs of the deck are covered with concrete. The footings probably represent the first use of concrete for structural purposes in an American bridge (see p. 158).

A sudden radical step—and for long, an isolated one—separates the conventional Roman arch from the first long-span segmental arch in the United States. When Montgomery C. Meigs planned the water supply system for Washington, D.C., which was constructed between 1857 and 1864, he was faced with the necessity of bringing the primary aqueduct over the deep ravine of Cabin John Creek. He boldly decided to bridge the stream with a single arch of 220-foot span and 57-foot rise, a low rise-span ratio that sprang from the precedent of Jean Rodolphe Perronet's daring bridges of the previous century. The long span of the Cabin John Aqueduct proved to be more economical to construct than a multiple-arch bridge of the common 50-foot span. The arch barrel is composed of dressed granite voussoirs, and the spandrel walls are thin strata of sandstone. An unusual feature of the sandstone slabs is that they are laid on radial lines for some distance above the upper surface of the arch barrel, presumably to increase the compressive strength of the arch, although this was probably unnecessary in view of the depth of the granite ring (4 feet at the crown and 6 feet at the springing).

Bridges in the form of the Cabin John Aqueduct were extremely rare before the advent of reinforced concrete, which quickly superseded the traditional masonry. Most bridges built after 1900 that appear to be stone are either concrete or steel structures with a stone facing added for ornamental effect. Builders began to follow this practice in the 1890's. The largest example is the third Pennsylvania Railroad bridge over the Susquehanna River at Rockville, Pennsylvania (1900–1902), although here the technique may have been adopted partly because of lack of confidence in the durability and weather-resistance of concrete. The masonry-arch bridge ceased to lead an active life chiefly because of its high cost—of quarrying the material as well as constructing the finished work—but also because the multiplicity of piers makes it inadmissable as a crossing over navigable waterways.

Chapter Six | Early Iron Construction

The most profound and far-reaching revolution in the history of the building arts accompanied the widespread adoption of iron as a structural material, which coincided in its pioneer phase with the Industrial Revolution of the late eighteenth century. There were two aspects of this technological reorientation, each causally related to the other in a complex pattern of interdependency. The immediately obvious one arose from the physical properties of the metal itself: iron is at least as strong as stone in compression, but unlike stone, it is equally strong in tension and shear; it is also an elastic material which can recover its shape after considerable deformation. Thus it has all the desirable properties of wood, but to a much higher degree, and has in addition the incombustibility of masonry. The ultimate consequence for structural techniques was that buildings and bridges could be very greatly increased in all their dimensions and radically altered in their form and at the same time their structural material could be considerably reduced in volume. With the greater size of buildings came greater openness of the interior and eventually the elimination of the external bearing walls. Before the end of the nineteenth century this was to work a revolution in architectural form.

The other aspect of the technological reorientation was not at first accompanied by any visible effect. The introduction of iron construction brought with it the necessary transformation of building from an empirical and pragmatic craft into a branch of scientific technology. This change began around 1750 and was one of the most pervasive economic and material consequences of the scientific revolution that occurred in the sixteenth and seventeenth centuries. This evolutionary process, bound up with a wide spectrum of cultural activities ranging from homely craft to philosophical speculation, forms a dramatic chapter in the history of science and technology, one that has yet to be treated adequately by historians. Without the development of a science of structure and materials, construction in iron and steel could never have been accomplished on the scale necessary to modern urban life. As the engineer was faced with constantly increasing demands for higher buildings and heavier, longer bridges, he was increasingly compelled to turn to science in order to solve the structural problems thrust upon him. Building could no longer be treated as an art or a craft; it had to become a branch of theoretical and applied science.

The use of iron as a structural material actually began in classical antiquity. Wrought-iron beams were placed in the architraves of large Greek temples to share the load of roof and upper courses with the stone lintels of the entablatures. But it is questionable whether the Romans ever used the metal, and the idea of iron structural members other than ties and cramps was lost during the Middle Ages in Europe. In the Orient one special use of iron seems to have had a continuous history from the early centuries of the Christian era. The Chinese were building suspension bridges with wrought-iron chains by the sixth century, but the form did not appear in Europe until the eighteenth, although iron tie rods and iron chains, as auxiliary structural members, were introduced in arcaded and domed structures during the Renaissance period. All iron used in Europe was wrought iron until the fourteenth century, when coke furnaces were developed that could generate temperatures sufficiently high to melt the metal into a wholly fluid state suitable for casting.

But the great period of iron construction—the Iron Age as it has been aptly called—begins with the Industrial Revolution of the eighteenth century. Perhaps the earliest systematic use of the material for primary structural elements came with the construction of St. Anne's Church in Liverpool, England (1770–72), the internal frame of which was composed of cast-iron columns and timber beams. The same technique was first applied to industrial buildings with the construction of the Calico Mill at Derby, England, designed by William Strutt and built from 1792 to 1793. At about the same time, a complete system of internal cast-iron framing combined with brick bearing walls was introduced by Charles Bage in his design of a flax mill at Shrewsbury, England (1796–97). By 1800 the internal iron frame was well launched in England, chiefly through the mills and factories built by Matthew Boulton and James Watt. But the aimless experimenting of many builders seeking to find the proper forms of iron members and the growing number of accidents caused by the collapse of iron structures made it clear that iron, for all its virtues, was little understood and that a scientific investigation of its properties and its behavior under load was a necessity.

The science of structural materials began with the publication of Galileo's *Dialogues on the Two New Sciences* (1638). The first part of this classic of modern mechanics represents the initial attempt to place the behavior of loaded structural members on a systematic empirical and theoretical basis. It contains many errors, but it provided the groundwork for subsequent progress, effected chiefly by Robert Hooke, Edmé Mariotte, Leonhard Euler, and Charles Coulomb in the seventeenth and eighteenth centuries. The great period of theoretical development came with the establishment in 1794 of the École Polytechnique at Paris. Before that, however, systematic experimental work on the strength of iron and other building materials had been initiated by Pieter van Musschenbroek at Leyden and the results published by him in tabular form in 1729. The decisive event in the progress of the new technology came in 1742: in that year Le P. Le Seur, François Jacquier, and Roger Boscowich made the first scientific attempt to calculate the dimensions of iron members when they were asked by the Vatican to investigate the defects in the dome of St. Peter's Church and to recommend corrective measures. The three men based their proposals on Musschenbroek's tables and on certain principles of theoretical statics formulated in embryonic form in the thirteenth century by Jordanus de Nemore. By the early years of the nineteenth century theory and experiment were close to the point where they could be brought to bear on the problems of building design.

The work of the brilliant French mathematicians needed to be supplemented by intensive investigations of the behavior of iron members under actual conditions of use. Here the practical English builders played their most valuable role. Charles Bage took the initial step in 1803, when he made the first full-scale load tests of iron roof frames. This was followed by a long series of experiments carried out by William Fairbairn and Eaton Hodgkinson at Manchester from 1826 to 1830. The results were recorded in a work entitled *Theoretical and Experimental Researches to Ascertain the Strength and Best Form for Iron Beams,* first published in 1831 by the Philosophical Society of Manchester and a year later serialized in the American *Journal of the Franklin Institute.* This work was followed by Hodgkinson's *Experimental Researches on . . . Cast Iron* (1846) and by Fairbairn's *Application of Cast and Wrought Iron to Building Purposes* (1854). The chief discovery of the English investigations was the realization that cast iron, which has a high compressive but relatively low tensile strength, is best suited to columns (compression members), whereas wrought iron, which is high in tensile strength, is best suited to beams (the members subject to bending and hence tension). The great misfortune—indeed, tragedy—is that American builders paid too little attention to these valuable and influential documents. Building in the United States up to the Civil War continued to be narrowly pragmatic, usually the work of men innocent of scientific training, with the result that failures of early iron bridges and building frames occurred with discouraging frequency.

The use of iron structural members for buildings was introduced in the United States by Benjamin Latrobe, but how much he and his immediate successors knew of the French and British work has never been ascertained. Latrobe was a practical builder of great ingenuity and responsibility who undoubtedly kept himself informed of the English achievements in iron-framed construction. These probably formed the background for his own initial essay, Christ Church in Washington, D.C., built in 1808. The second use was the Chestnut Street Theater in Philadelphia, designed by William Strickland and built between 1820 and 1822. The two balconies and ceiling of the theater rested on three stands of cast-iron columns, which were disposed in a semicircle to conform to the plan of the balconies, the columns of each stand extending in line one above the other from the foundation to the domed ceiling. The columns were fixed at the base in iron sockets set in independent stone footings. The theater was demolished in 1856, and because none of the surviving drawings shows any details of the interior structure, we have no idea of the form and size of the columns. They were undoubtedly hollow cylindrical members, since this type had already been recognized by Bage and his English contemporaries as having the most efficient form.

The third American use of iron columns was also Strickland's work. The United States Naval Asylum at Philadelphia, built between 1826 and 1830, consists of a central block and two long flanking wings. The construction is chiefly masonry, with brick floors resting on vaulted brick ceilings. Two iron-floored galleries or piazzas under an overhanging roof run the length of the wings on both the front and rear. The lower gallery is supported on granite posts of square section, but the upper one and the roof overhang are carried on a total of forty slender cast-iron columns about 8 inches in diameter and spaced about 14 feet on centers. The Naval Home still stands, and although somewhat altered from its original appearance, its essential structural elements remain unchanged.

These early essays in Philadelphia prepared the way for an extraordinary work of iron framing erected in the city of its origin when the new technique was scarcely known in New York and Boston. The Philadelphia Gas Works was granted a permit in 1835 to build a generating plant and storage tanks on Market Street at the Schuylkill River. The first installation was completed probably in 1837 and substantially enlarged in 1850. The dangerously inflammable product required the most extensive fireproof precautions available at the time. The retort houses were constructed with brick walls, the floors and retorts resting on brick arches spanning between cast-iron girders. The roof frames were described in an early Philadelphia guide as wrought iron. The description apparently applies to the 1837 building, which thus marks the first use of wrought iron for structural purposes in the United States. No description of the gas holders has survived, but prints suggest that the tanks were built up of cast-iron plate and the frames of cast-iron columns and girders. The three holders of the 1850 addition were much larger, and the guide frame in each case consisted of a ring of cast-iron columns arranged in stands one above the other and tied at the top of each level by iron beams cast as classical entablatures, except for those of the topmost ring, which were joined by shallow trusses. The one remaining illustration of these structures shows no bracing other than the ring of trusses, which may have been bolted to the columns throughout the depth of the truss web. Among the valuable lessons embodied in these iron frames was perhaps the first example of portal bracing in a structure other than a bridge (for his type of wind bracing, see p. 123). The first systematic use of portal bracing in a building came with Joseph Paxton's Crystal Palace, constructed for the London Exhibition of 1851; the technique does not appear in an American building other than the New York Crystal Palace of 1853, until 1890.

In the development of iron framing New York at first lagged behind Philadelphia, but the larger city was soon to establish a preeminence in iron construction that was not to be challenged until the rise of the Chicago school in the 1880's. Philadelphia might have maintained its lead had it not been for the compelling factor of disastrous fire in New York. On December 16–17, 1835, seventeen blocks of buildings on the lower East Side were destroyed by fire, and the catastrophe was to repeat itself ten years later in roughly the same area. The requirements of fireproof construction and strength to support warehouse loads forced the builder to turn to iron in place of the vulnerable timber framing. The first works were hesitant and sporadic. The beginning was apparently Andrew Jackson Davis's Lyceum of Natural History, built in 1835 on Broadway near Spring Street: iron columns at the first story of the façade supported the brick wall above. A similar system of construction was used two years later in the Lorillard Building on Gold Street, where three square-section iron columns in the façade extended through the first two stories to carry the brick wall of the three above. In 1846 John Travers introduced iron beams into a building when he built his fireproof library at Paterson, New Jersey.

These little structures constitute the local basis for the work of the two most ambitious and influential manufacturers of iron building components in the nineteenth century. James Bogardus and Daniel Badger turned the casting of beams and columns into a major industry, and they were involved in the construction of buildings that ultimately had a revolutionary effect on American structural techniques. Bogardus was the more imaginative and scientifically oriented of the two. Born in Catskill, New York, in 1800, a prolific inventor and lecturer on technical subjects, he learned much of his art from a European visit extending from 1836 to 1840 that he made expressly for the purpose of studying iron construction. In 1848 he established a foundry at Centre and Duane streets in New York to cast columns and beams as well as the iron parts of a sugar mill he had invented. The foundry itself was the embodiment of Bogardus's theory of iron construction, which was set forth in detail in the drawings he prepared for his patent of 1849 (Fig. 24). His original intention was to construct a building composed entirely of cast-iron elements, but this was realized only in part. His broader aim was both functional and aesthetic: to create a durable, economical, and fireproof structure and to imitate as cheaply as possible the forms of classical architecture.

The extreme simplicity of the foundry construction brought it as close to the mechanization of building techniques as was possible at the time. At the base was a continuous footing of stone extending around the periphery of the building and supporting a cast-iron sill planed to a smooth surface and uniform thickness. Hollow cylindrical columns turned on the lathe, with flanged ends machined smooth, were bolted to the sills through the column flanges at the sill joints. Spandrel girders in the form of channels were bolted to the top flanges of the columns. Another set of sills, columns, and channels was added for each succeeding story. The outer members of the iron frame took the place of a bearing wall, which then had no function and was omitted. The floors of the factory were carried on timber beams, but in all other respects the structure was simply a cast-iron cage. Bogardus designed it so that it could be disassembled by drawing the bolts and re-erected on another site. The disassembly was accomplished in 1859, but there is no evidence that the factory was ever rebuilt. By that date Bogardus's business had far outgrown its original quarters. In the patent drawings of 1849 he proposed a floor and roof construction of cast-iron plates with tongue-and-groove joints, floor girders in the form of shallow segmental arches supplemented by wrought-iron tie rods, and floor and roof beams of I-section. The now universal I-beam was introduced in the United States by Bogardus, who probably learned of it through the publication of the Fairbairn and Hodgkinson works. The great virtues of the

23. Starrucca Viaduct, New York and Erie Railroad, Susquehanna, Pa., 1847–48. John P. Kirkwood, engineer. This high and daringly proportioned bridge still carries the trains of what is now the Erie-Lackawanna Railroad.

24. James Bogardus factory, Centre and Duane streets, New York City, 1848–49. A pioneer work of cast-iron building construction in the United States.

25. Haughwout Store, Broadway and Broome Street, New York City, 1857. John P. Gaynor, architect. The little building still survives as a classic of the cast-iron front.

26. Wanamaker Store, Broadway at Ninth Street, New York City, 1859–68. John Kellum, architect. The iron frame around the interior light court, exposed after the fire of July, 1956.

27. Court House, St. Louis, Mo., 1839–64. Singleton, Twombley, Mitchell, Lanham, and Rumbold, successively, architects. Section through the dome. Since the St. Louis Court House antedated the dome of the United States Capitol by two years, it proved to have the first cast-iron dome in the United States.

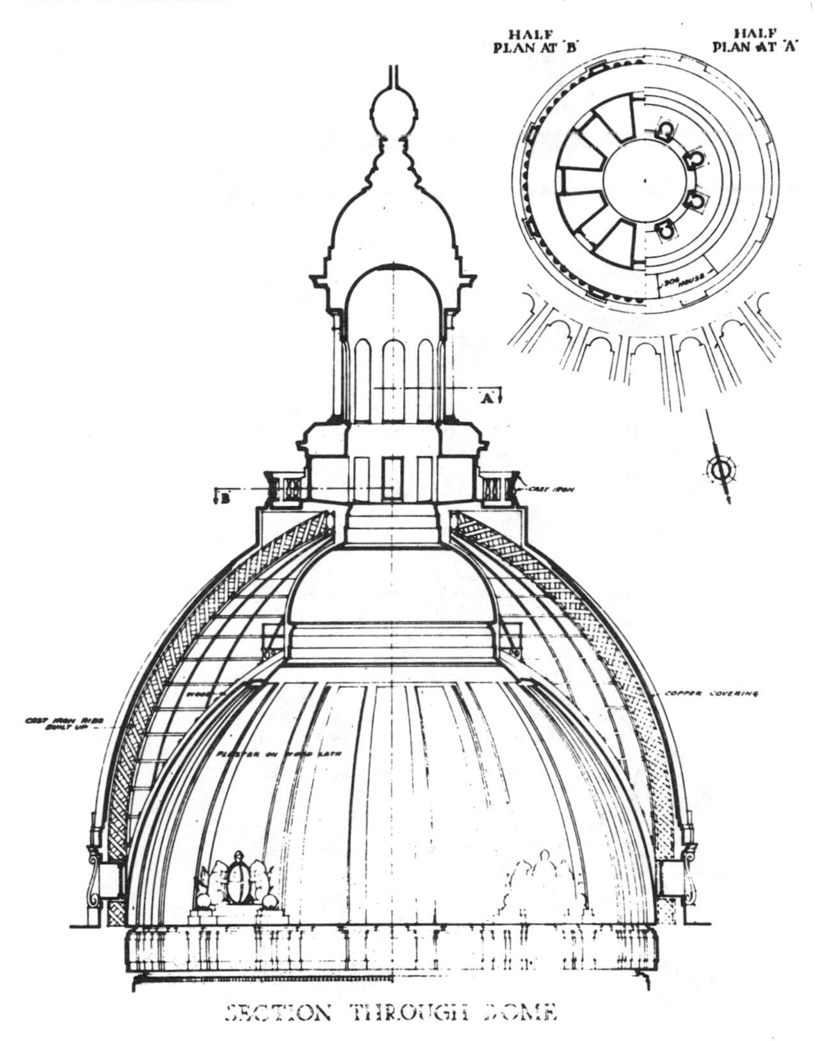

28. United States Capitol, Washington, D. C., 1856–64. Thomas U. Walter, architect. Section through the dome. Only a sectional drawing reveals the complex truss construction that supports the iron ribs.

29. Washington Monument, Washington, D. C., 1833, 1848–84. Robert Mills, architect. Top of the stair and elevator shaft during construction, showing the Phoenix columns.

30. Iron truss patented by August Canfield, 1833. The pioneer form was an odd hybrid, part suspension bridge and part truss in its action.
31. Trusses patented by Thomas and Caleb Pratt, 1844. Because they are the most efficient forms, the Pratt and Warren trusses are now nearly universal in steel truss-bridge construction.

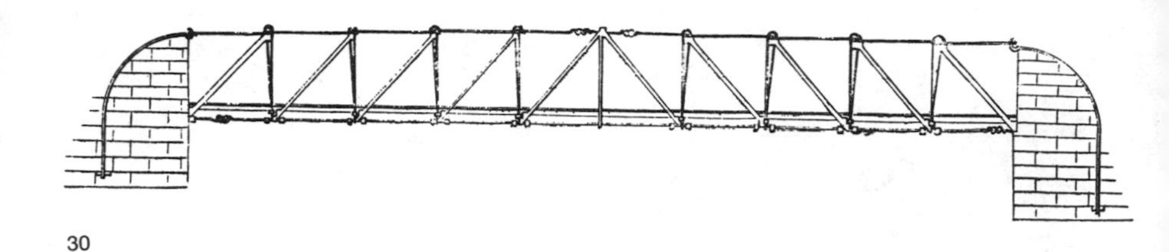

30

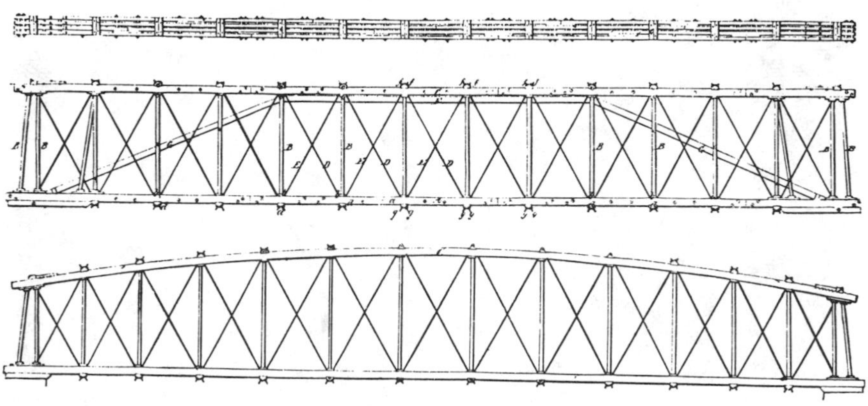

31

32. Forms of the Warren truss. (*a.*) The original form patented by James Warren and Theobald Monzani, 1848. (*b,c.*) Later modifications.
33. Bridge of the Baltimore and Ohio Railroad, Potomac River, Harpers Ferry, W. Va., 1862–63, 1868. Bollman truss spans.

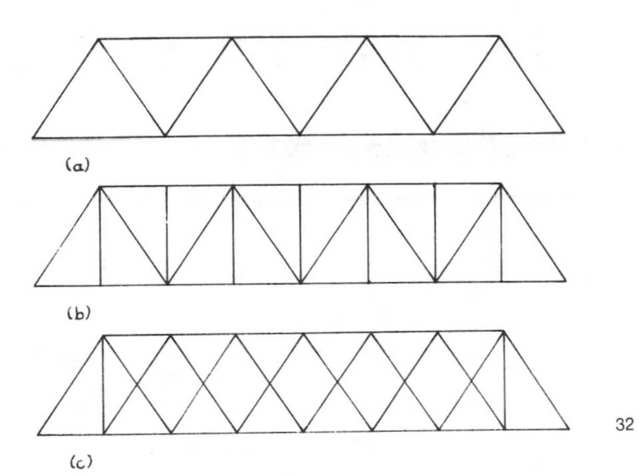

33

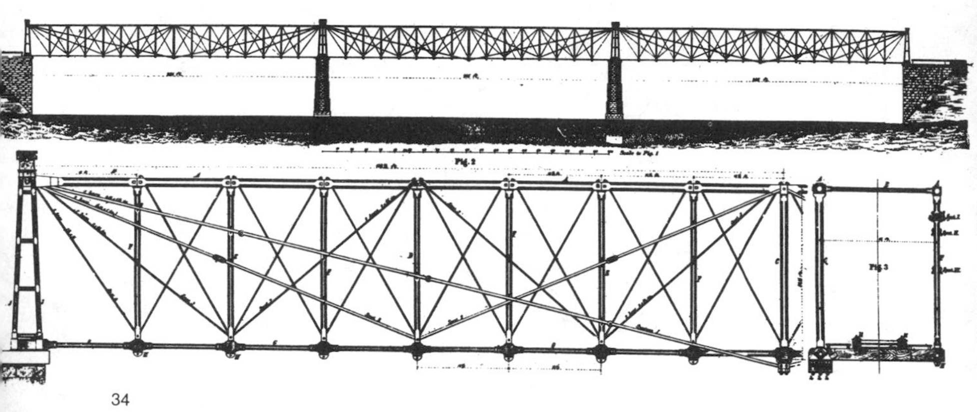

34

34. Bridge of the Baltimore and Ohio Railroad, Monongahela River, Fairmount, W. Va., 1851–52. Albert Fink, engineer. An early example of the through truss patented by Albert Fink.

35

35. Bridge of the Union Pacific Railroad, Missouri River, Omaha, Neb., 1868–72. Post truss spans.

36. Street bridge, Schuylkill River, Philadelphia, Pa., 1841–42. Charles Ellet, engineer. An early wire-cable suspension bridge, by the man who preceded Roebling in its American development.

37. Suspension bridge, Ohio River, Cincinnati, Ohio, 1856–67. John Roebling, engineer. The bridge as it was originally constructed. The second largest of the Roebling spans, the strengthened bridge still carries a heavy vehicular traffic between Cincinnati and Covington, Kentucky.

38. Produce Exchange, New York City, 1881–84. George B. Post, architect. Cross section showing the interior framing. The Produce Exchange came very close to being full skeletal construction.

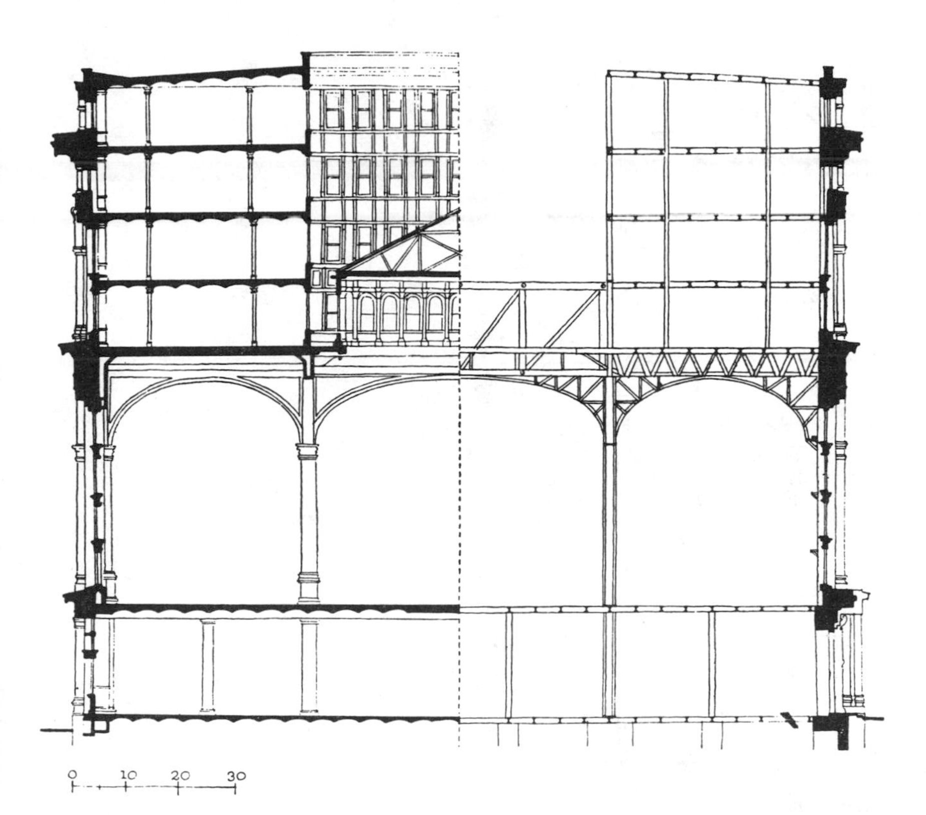

39. Havemeyer Building, New York City, 1891–92. George B. Post, architect. Cross section showing the interior frame. The sectional drawing reveals the New York builder's continuing dependence on iron and masonry construction even for very high buildings.

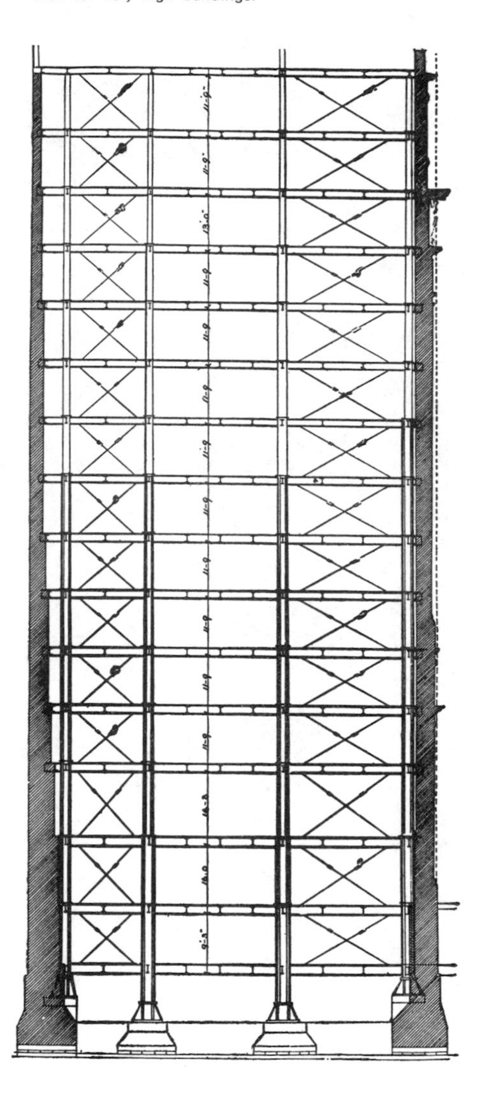

40. Auditorium Building, Chicago, Ill., 1887–89. Adler and Sullivan, architects. Longitudinal section. The interior framework of the Auditorium embraced every structural technique in iron available at the time.

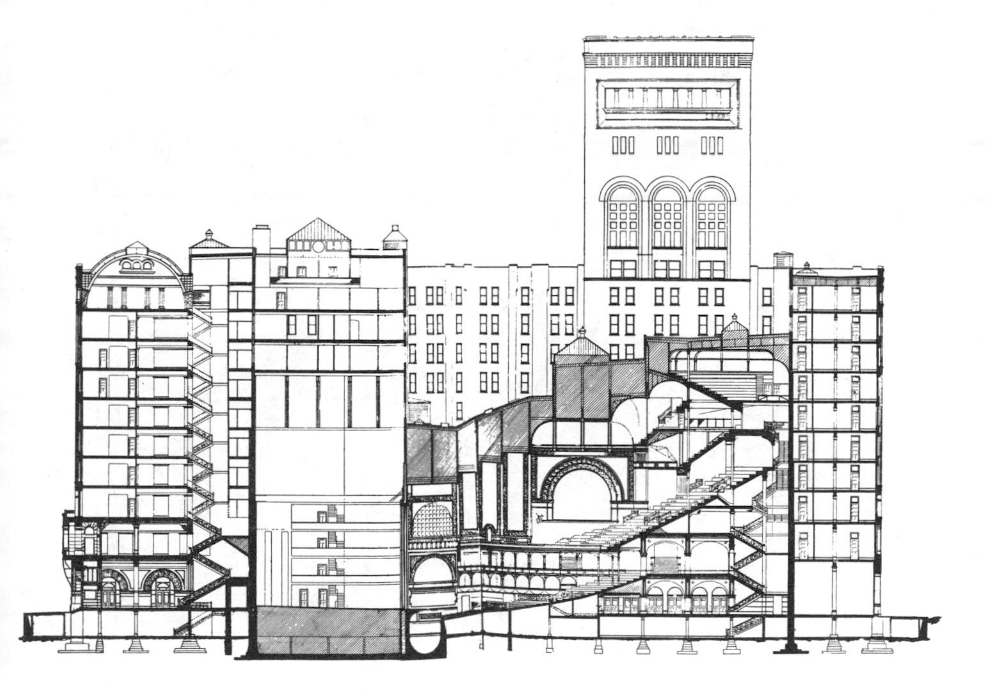

41. Monadnock Building, Chicago, Ill., 1889–91. Burnham and Root, architects. Longitudinal section. The heavy masonry walls were already out of date in Chicago by the time John Wellborn Root designed this architectural landmark, but the interior iron frame of the Monadnock embodied the most advanced principles.

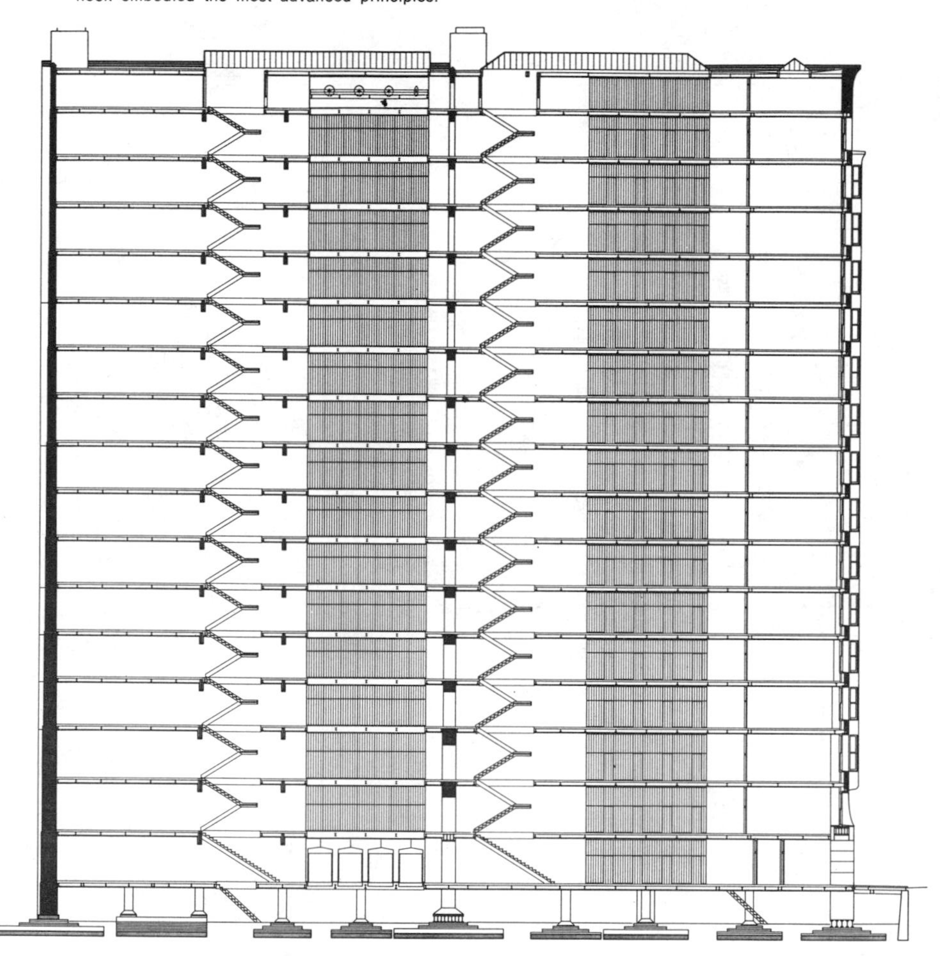

Bogardus system were offset by a few defects, although these were not of a radical nature for small buildings: the lack of rigidity of the hand-bolted unbraced frame, the poor fire resistance of exposed iron, and its vulnerability to corrosion in acid-bearing city air.

The first large commercial building erected by Bogardus was the second printing plant of Harper and Brothers, built to replace the original facilities, which were destroyed on December 10, 1853, in the first million-dollar fire in the United States. The new building was designed by John B. Corlies and constructed on Franklin Square, New York City, in 1854. Arched cast-iron floor girders with wrought-iron ties were supported by cast-iron columns and brick partitions. The structural innovation of greatest importance was the transmission of floor loads to the girders by means of 7-inch wrought-iron rail beams, part of the first lot manufactured in the United States and expressly designed for the new wide-bayed iron framing. The lot was rolled in the spring of 1854 at the Trenton Iron Works, founded by Peter Cooper and Abram Hewitt, and the various beams installed almost simultaneously in the Harper plant, the New York Assay Office, The Cooper Union, and a structure built for the Camden and Amboy Railroad. Smaller wrought-iron beams and rails had been rolled since 1784, when the Englishman Henry Cort invented the grooved roller and the puddling process that made their manufacture possible; the problem had been to develop a mill capable of rolling 7-inch beams, which was finally solved by Cooper's engineering staff at the Trenton works.

Bogardus's most prophetic work for the evolution of structural techniques was the shot tower built for the McCullough Shot and Lead Company in New York City in 1855. The tower was 175 feet high and was supported entirely by an octagonal cast-iron frame that formed a true system of skeletal construction. The frame consisted of eight inward-leaning stands of cast-iron columns that were tied at eight levels by peripheral cast-iron beams. The panels formed by adjacent pairs of columns and beams were closed with brick screens that rested on the beams and thus constituted a non-bearing or curtain wall. There is reasonable evidence that this structure was known to William Le Baron Jenney, the Chicago architect who made the greatest contribution to the development of the skyscraper in his Home Insurance Building in Chicago (see pp. 123–25).

Daniel Badger, Bogardus's chief competitor, began his career in Boston in 1829 and established his immensely prosperous Architectural Iron Works at Duane and Thirteenth streets, New York, in 1847. While Bogardus tended to be sober and analytical in his presentation of the virtues of cast iron, Badger expressed unbounded enthusiasm for the new material. He recommended it for the structural and ornamental details of every kind of utilitarian building and claimed that it had no equal for every virtue asked of a building material. His rather exaggerated claims quickly bore fruit: within a few years he was casting iron structural members for an extraordinary number of buildings that were multiplying on the streets of New York, Chicago, and other major cities, including Havana, Cuba.

The classic example of his work is the store building designed by John P. Gaynor and erected in 1857 for E. V. Haughwout and Company at Broadway and Broome Street (Fig. 25). In this building Elisha Graves Otis installed the first passenger elevator with an automatic brake. The store is typical of Badger's commercial structures, which ranged in height from two to six stories that were clearly defined by the uniform rows of iron columns and spandrels that appeared in the street elevations. The columns were usually closely spaced, seldom over 6 feet, and the area between them entirely opened to glass. The interior frames of the buildings bore little relation to the rhythmic cast-iron façades. Here hollow cylindrical cast-iron columns were usually combined with timber beams and joists, or less often, wrought iron, and the bay span was two or three times longer than the column spacing in the street elevations. The important exception to the general run of commercial buildings was the grain elevator constructed for the United States Warehousing Company in Brooklyn, New York, in 1860 or 1861. The designer was George H. Johnson, who was employed by Badger at the time and who later founded a business to manufacture the fireproof tile that he patented in 1871. The construction of the elevator followed the precedent of Bogardus's McCullough shot tower: the spandrel girders of the all-iron frame carried the panels of the brick wall. which again were reduced to nothing more than a protective screen. On two trips Johnson made to Chicago to promote the sale of his fireproof tile, he may have provided Jenney with information about the Brooklyn elevator, which may thus have become another precedent for the Home Insurance Building.

The high point of iron framing for the size of the total enclosure and the variety of structural forms came with the Crystal Palace, designed by the Danish-born architects George Carstensen and Charles Gildemeister for the New York Exposition of 1853. Although inferior in architectural design and structural efficiency to its immediate predecessor and namesake, the Crystal Palace of the London Exhibition of 1851, the New York building was an impressive demonstration of the versatility of iron construction. Cast-iron columns supported an elaborate portal-braced system of wrought-iron arch ribs, trussed girders under the flat and gable roofs, and arched trusses under the great central dome. The whole iron structure was strongly etched against the curtain walls of glass. The framing of the dome was an important if ephemeral forerunner of the structural systems of monumental domes like those of the Capitol and the balloon train sheds of the big railway terminals.

The masterpiece of iron construction for commercial purposes was the Wanamaker Department Store that stood for nearly a century on lower Broadway between Ninth and Tenth streets in New York City (Fig. 26). Originally built by Alex T. Stewart, one of the founders of the department store business, it was erected in stages between 1859 and 1868 after the plans of John Kellum. The manufacturer of the structural and ornamental iron was the Cornell Iron Works of New York. The five stories of the Wanamaker Store embraced a total floor area of 325,000 square feet, sufficient to make it the largest iron building of the time other than the temporary Crystal Palace. The street elevations were open patterns of slender, closely ranked iron columns and narrow spandrels extending continuously along the length of the walls. The interior construction consisted of plank floors and timber joists spanning between wrought-iron beams that rested in turn on cylindrical cast-iron columns. The framing at the periphery of the central light court was more massive:

deep wrought-iron girders extending around the court were carried on columns of rectangular section; above the opening was a gable skylight supported by triangular wrought-iron trusses. All girders and beams were fixed to shoulders cast integral with the body of the column by bolting the lower flange of the beam to the shoulder. This was far from a rigid system of connections, and the absence of bracing exposed the building to possible damage from the bending and twisting action of wind. The preventive factors were the great weight of cast iron in the elevations and the relatively extensive horizontal dimensions for the height. The Wanamaker Store was struck down by a spectacular fire on July 14–15, 1956. The flooring, joists, and other combustible elements were totally consumed, leaving the naked frame to stand free and undamaged. The wreckers finished the famous landmark in the next few weeks.

Iron-fronted and iron-framed buildings continued to be erected up to the eighties, but after the Civil War their popularity began to decline. The structural system developed by Bogardus contained valuable features for the subsequent evolution of building techniques: it demonstrated the strength, economy, and durability of the iron frame, and at the same time it made possible an open interior in which light movable partitions could be arranged to suit the needs of the owner. The iron-framed exterior walls were in part assemblies of prefabricated elements and thus proved to be the forerunner of prefabricated curtain-wall units, although a long hiatus separated the iron fronts of the Civil War era from the manufactured wall elements of the contemporary skyscraper. But the bolted iron frame had serious limitations for the high-rise buildings that were soon to come in New York and Chicago. For these, the braced and riveted frame was a necessity. Ironically enough, the New York builders returned to the masonry bearing wall for the skyscrapers that multiplied after the Civil War, and they clung to it for nearly thirty years before they finally accepted the fully developed iron and steel skeleton. The revolution came in Chicago, where the iron front never reached the level it achieved in the East.

Chapter Seven | Monumental Buildings in Masonry and Iron

The application of iron to particular parts of monumental masonry buildings forms a separate chapter in the evolution of American building techniques. The technique came into existence to deal with the exigencies posed by a special kind of building, and its history is short, falling entirely within a few decades around the time of the Civil War. The construction of huge monumental structures like the United States Capitol involved peculiar problems arising from the domed enclosures: full masonry construction would have been far too costly and unmanageable in weight and scale, and concrete—the Roman solution—was insufficiently developed to meet needs of such magnitude. The only alternative was iron, but it had to be used in special ways for which there was little precedent, and much of it had to be hidden so as not to detract from an appearance established by the canons of Renaissance and Baroque architecture. The chief historical importance of the techniques we are considering here lies in their role in furthering the progress of iron construction, since traditional masonry eventually reached a dead end.

The problem of building and supporting large ribbed domes with a rise greater than their radius offered a formidable challenge to the builder. The first successful solution in iron came with the construction of the old Court House at St. Louis, now expertly restored and preserved as part of the Jefferson Expansion Memorial Park. The talents of five architects and the labor of twenty-five years went into the production of this building. The cornerstone was laid on October 21, 1839; the rotunda, the original dome, and the east-west wings were completed by 1852; the north and south wings and the present dome were completed in 1862, but further changes in details delayed the final opening until 1864. The five architects who had a hand in the design were, successively, Henry Singleton, William Twombley, Robert S. Mitchell, Thomas D. P. Lanham, and William Rumbold. Singleton and Twombley found the job too difficult and resigned a year after they had been appointed; Mitchell was largely responsible for the main enclosure as it now stands; Lanham played a minor role in the design of the rotunda and the north wing; and Rumbold, who possessed the most original constructive talent of the five, designed the iron dome. He was granted a patent on the structural system in June, 1862.

There was good reason for the presence of a major work of iron building in St. Louis. The region enjoyed a plentiful supply of iron ore at Iron Mountain, Missouri, sixty-five miles south of the city. Philip Renault had established a local iron industry at least twenty years before the city was founded in 1764. By 1800 St. Louis had become the focal point of the whole trans-Mississippi trade, and the flourishing steamboat traffic pushed the city's prosperity to steadily rising levels in the first half of the nineteenth century. As in the case of New York, fire was the chief stimulus to iron building construction. The St. Louis disaster came in 1849, so that by the mid-century decade iron fronts and internal iron frames were rapidly multiplying among the commercial buildings of the waterfront. The expanding population of St. Louis required a large court house early in the nineteenth century, and the citizens were willing to pay for one that would express the wealth and importance of the city.

The Court House as completed is very nearly an equal-armed cross in plan, with an over-all length along the north-south wings of 229 feet. The dome rises at the crossing of the wings to a height of about 230 feet above grade. The mass of the building is masonry construction throughout: the walls of the east-west wings are brick with limestone facing on the east (front) elevation; the columns of the entrance portico are limestone, as are the walls of the newer north-south wings, but the roofs of all the wings are supported by wrought-iron trusses. The rotunda columns rest on a ring footing probably of limestone, although this has not been established exactly. The area under the flagstone paving of the rotunda floor was not excavated, so that the stone slabs rest directly on the earth. The rotunda columns are hollow cast-iron members, but those supporting the circular galleries of the third and fourth floors are both wood and iron, the former resting on radial girders, the latter bearing on the ring of iron columns at the rotunda level.

The ceilings of the basement corridors are groin vaults of brick sustained at their corners by buttresses in the brick partitions. The ceiling over the office area of the east wing, on the other hand, is composed of parallel brick arches spanning between iron beams supported by the interior brick partitions and the outer foundation walls. The beams span 20 feet 4½ inches, a remarkable length for iron beams cast in 1848, when the east-west wings were completed. If these members are wrought iron, as has been claimed, they must have been added after the wing was completed; as a matter of fact, even cast-iron beams of 20-foot length would have been unlikely in St. Louis at that early date. Another unusual detail in the basement construction is the brick vault that supports the entrance stairway. The span of the vault is equal to the length of the stairway, and the plane between the springing edges lies on a slope parallel to the central axis of the stairs.

The high dome of the St. Louis Court House clearly marks a major innovation in American building practice. The entire structure from bottom to top is an unusually complex example of construction in several different materials. A circular masonry wall on an independent footing, located outside the ring of rotunda columns, carries an octagonal masonry wall that extends above the roof level. This wall, which changes to a smooth cylinder above the roof level, supports a cast-iron drum with engaged columns. On this rests the double dome with an ornamental cupola at its crown (Fig. 27). The interior surface of the inner dome is plaster on wood lath; the outer surface is covered with wood sheathing on meridional timber ribs. The outer dome is a cast-iron shell carried on a series of meridional trusses in the form of a Howe truss without radial members, that is, members equivalent to the vertical rods of the flat bridge truss. The horizontal thrust of these trusses is taken by the iron columns of the drum, and at the crown they bear on a circular opening acting as a compression ring. The trusses are tied transversely by rings of iron strap. The ironwork of the dome and drum was meant to function as a unit, but its action is separate from that of the masonry structure that supports it.

One other detail of the Court House provides further evidence of the ingenuity that went into its creation. The main interior stairway, installed in 1854, is cast iron in the form of a compressed or oval helix that rises 32 feet from the main to the third floor. Except for one supporting post located at the outer edge of the bottom step, this stairway is cantilevered from the masonry wall at its outer periphery. A curious feature is that the extension of the main floor in the stairwell is composed of three heavy limestone slabs also cantilevered from the wall, a rare phenomenon in stone. All the stairways in the building are self-supporting iron flights, but except for the one we have described, they are made up of straight runs. The many technical innovations in the Court House place it in the front rank of structures of its kind, but it is questionable whether it bears any organic relation to subsequent developments other than the repetition of the monumental dome in later governmental buildings.

The dome of the Capitol in Washington is a much larger structure and a more influential one because of its popularity. It was completed two years after the one at St. Louis, but the designing and construction of the two were carried out independently of each other. President Millard Fillmore appointed Thomas U. Walter architect of the Capitol in 1851. The chief engineering consultant was the bridge and hydraulic engineer Montgomery C. Meigs, and his principal assistant was August Schoenborn. The architect's first task was the design of the House and Senate wings, which were constructed between 1852 and 1854. Together with the central area, these extend for an over-all length of 751 feet 4 inches. The walls are traditional bearing members of masonry, but the roofs are carried on triangular wrought-iron trusses. To complete the fireproofing of the structure and the permanent fixtures, Walter specified cast iron for the ceiling panels, window frames, moldings, and trim. This marked the most extensive use of cast and wrought iron in any American building erected up to that date.

The great structural and visual feature of the Capitol is the dome, which rises to an over-all height of 287 feet 5 inches above the main floor (Fig. 28). Its design and construction constituted a daring enterprise at the time because of the magnitude of the operation and the lack of precedent to guide the designers. The only completed iron dome in existence at the time was that of the Cathedral of St. Isaac in St. Petersburg (now Leningrad), Russia. Walter and his engineering associates studied it thoroughly before they undertook the design of their own project in the mid-1850's. The plans were probably prepared by 1856, but construction dragged on until 1864 because of piecemeal appropriations and the exigencies of war. Walter first proposed that the shell be supported by a system of meridional trusses with straight inner chords and elliptical outer chords conforming to the curve of the dome section. This scheme was abandoned mainly for two reasons: the few web members would have resulted in inadequate rigidity, and the great depth of the straight-chord truss at the haunch made the proper curvature and size of the inner shell an awkward if not insoluble problem. Accordingly, the more sophisticated proposal advanced by Schoenborn was adopted.

The dome as built is a cast-iron structure throughout. The chief supporting elements of the ribs are meridional iron trusses in which both outer and inner chords are elliptical, though with a gentler curvature of the dome section. These trusses have a dense array of web members disposed in a series of X's, like the diagonals of a Howe truss. The meridional trusses are carried to a ring footing of solid masonry by a circle of buttresses marked by pilasters on the rotunda wall. This footing was originally built to support the old timber-framed dome designed by William Thornton. The weight of the iron ribs in the dome is in part carried to the trusses by a system of radial posts braced by single diagonals between pairs of adjacent members. The drum and cupola above the eye of the dome are built up of iron posts and ribs. Two inner shells constitute the double ceiling over the rotunda: the upper one is smooth plaster painted with various allegorical scenes; the lower, a truncated hemisphere in shape, is finished in imitation of coffered masonry. Typical of Baroque domes, these inner shells are entirely separated from the iron structure above them.

A unique work of iron and masonry construction in the United States is Washington Monument, although here the two materials form two separate rather than supplementary systems. The monument was designed by Robert Mills in 1833 and erected over a period of thirty-six years, from 1848 to 1884, because of the intermittent flow of private and state contributions that paid for its construction. The finished obelisk is a solid-walled skyscraper of marble 555 feet high, topped by a pyramid that terminates in a sharp-pointed aluminum cap. The form is perfectly stable against wind loads because of the battered walls and the great weight of masonry. Erecting the stone shaft was a matter of patience rather than constructive skill, and there was nothing unfamiliar about its form. The unique feature is the iron framework that defines the interior elevator shaft and acts as the support for the elevator system and partly for the iron stairway that surrounds it (Fig. 29).

The primary members of this frame are wrought-iron columns of the type known as Phoenix columns. The invention has been credited to Samuel Reeves, an engineer with the Phoenix Iron Company at Phoenixville, Pennsylvania, who was granted a patent on it in 1862. There is now evidence, however, that the column was originally developed by Wendel Bollman, a bridge engineer with the Baltimore and Ohio Railroad, for the supporting bents of a bridge erected at Havana, Cuba, about a year before the date of the Reeves patent. The column was a built-up wrought-iron member consisting of four or eight flanged longitudinal segments like straight barrel staves that were bolted together through the flanges into a cylindrical form. The main intention of the inventor was to design a column that would be easy to roll yet large enough to carry the heavy loads of bridges and high buildings and to withstand the bending forces arising from lateral loads and the buckling tendency of long columns. The flanges helped in the latter respect by increasing the rigidity of the member against bending.

The Phoenix columns of Washington Monument form the highest iron frame erected up to that time. The eight stands of columns are arranged in two groups of four each in the pattern of a square within a square. They are connected at the joints by peripheral girders, and the whole system is braced by diagonal members set at intervals in horizontal planes up the height of the tower and located between the corners of the two squares. The high tower-like frame inside Washington Monument grew out of a particular problem that the builder was not likely to encounter again, but it provided a convincing demonstration of the potentialities of the new technique for high building construction.

Chapter Eight | Iron Bridges

The requirements that led to the application of iron to building frames also dictated its choice for bridge construction. Among the various types for which iron is appropriate, the truss bridge was most closely associated with the growth of the rail network. The railroad builders were early compelled to turn to the new material because of their overriding need for structures with maximum strength, rigidity, and fire-resistance for a given quantity of material. Iron is an elastic substance, like wood, and is thus able to resist tension and shear as well as compression, but it is enormously greater in strength (18,000 pounds per square inch *minimum* tensile strength for ordinary cast iron, as opposed to 2,550 pounds *maximum* for common structural wood). Although we are astonished at how successfully timber could be used in large rail bridges, the day was rapidly coming when iron was to supplant wood for all but the smallest spans. The rapid increase in the weight and speed of locomotives meant that the bridge builder would have to rely on the new science of structure to a far greater degree than the architect of commercial buildings; yet in spite of what should have been an obvious fact, he sometimes followed the old pragmatic approach to the point where disaster finally forced him to come to terms with progress.

During the first decade of railroad building iron construction was impractical for a variety of reasons: there were few foundries and mills capable of producing iron members of the size necessary for bridge trusses; the weight of locomotives was small enough to be safely carried on timber spans; and the short lines that first reached out from Boston, Albany, New York, and Baltimore could seldom command the financial resources necessary to construct iron bridges. These resources came in the decade of the forties, simultaneously with the introduction of iron into the frames of buildings. And in another parallel to the development of building construction, before the mid-century decade, all-iron bridges were rare, combinations of timber and iron being the more common form. Regional factors introduced determinants other than those of traffic and financial resources: in the East, where there was an established iron industry close to the major rail centers, iron soon became common; in the

West, where its cost was prohibitive, timber remained dominant during the construction of the Pacific railroads, from 1862 to 1910.

The first bridge composed of iron throughout is the arch span built by Abraham Darby III between 1775 and 1779 to cross the River Severn at Coalbrookdale, England. Much larger and more daring structures were to follow in England and on the Continent before the end of the century, but more than fifty years were to pass before one appeared in the United States. The little span that inaugurated the native development of iron and steel bridges was constructed at Brownsville, Pennsylvania, 1836–39. The designer was Richard Delafield of the Army Engineers, and the manufacturer of the castings was John Snowden, the owner of a local foundry. An arch bridge of traditional form, its plank deck rested on stringers and beams probably in the shape of channels, and these in turn were supported through a spandrel frame by five hollow cast-iron segmental tubes built up of short lengths bolted together through circumferential flanges. The bridge survived unchanged until 1921, when samples of the iron from the original structure were analyzed by metallurgists of the American Rolling Mill Company at Middletown, Ohio—a simple and inexpensive practice of great value to the historian of technology but unfortunately a rare event in the United States. The analysis revealed that the purity of the iron was nearly equal to that of commercial rust-resistant iron manufactured in 1920, but it told nothing about the homogeneity of the metal, the lack of which was a frequent cause of structural failure in the nineteenth century. After the Brownsville span the iron arch was not repeated for more than twenty years. The second, built by Montgomery C. Meigs in 1858 to carry Pennsylvania Avenue over Rock Creek in Washington, D.C., also rested on arches in the form of hollow cast-iron tubes, but in this case they served the additional function of water mains.

The iron truss came soon after the iron arch, although the concept antedates the Brownsville span. As in the case of its timber counterpart, the practical iron truss was initially a creation of American builders rather than European. In 1833 August Canfield of Paterson, New Jersey, was granted a patent for an odd hybrid truss, half suspension and half-truss in its action (Fig. 30). The top and bottom chords and the vertical hangers were to be wrought-iron rods, and the diagonals rigid cast-iron members. The top chord, however, which was to be anchored at its ends in masonry abutments and from which the deck and truss were to hang, was apparently to act like the cable of a suspension bridge. The diagonal members were probably regarded as stiffening elements, like the stiffening trusses of modern suspension bridges, although they may have taken part of the load. It is impossible to make an accurate stress analysis of the Canfield truss as designed, although it could be analyzed if the structure were built throughout as a rigid truss.

There is no record that bridges of the Canfield design were ever built. The first constructed iron truss was a 77-foot span erected in 1840 to carry a road over the Erie Canal at Frankford, New York. The designer was Earl Trumbull, who received a patent on the form in 1841. The truss was essentially a double-diagonal Howe truss with all members cast in the once common form of an equilateral cross in section, or a +-section, as it was once designated. The bridge was braced laterally against wind loads by means of diagonal wrought-iron rods set between the beams and stringers of the deck frame.

Trumbull's bridge ushered in the prolific decade of the 1840's, when the dominant truss forms of the nineteenth century and the two that are universally used today were invented and given their first practical demonstrations. The construction of more than six thousand miles of railroad line provided the major impetus. The first iron bridge designed for railroad use was a girder span constructed by the Baltimore and Ohio at Laurel, Maryland, in 1839. The built up I-section girder, spanning 30 feet, was supplemented by a series of wrought-iron tension bars bowed downward by means of struts below the bottom flange. The first iron railroad truss appeared in 1845, when Richard Osborne of Philadelphia built a bridge for the Philadelphia and Reading Railroad at Manayunk, Pennsylvania. Three parallel double-diagonal Howe trusses carried a double-track line over a clear span of 34 feet. Following the division of stress in the typical Howe truss, the verticals were treated as tension members and were hence wrought iron, and the diagonals as compression members, were cast iron. But Osborne seems to have misunderstood the action of the chords. Both the top and bottom members were wrought iron: this is sound enough in the case of the bottom chord, which is subject to tension, but questionable for the top chord, which is in compression. One truss of the Manayunk bridge was salvaged during demolition and has been preserved in the Smithsonian Institution.

The third iron railroad bridge introduced another basic structural type in the United States: the second girder bridge was built between 1846 and 1847 by James Millholland for the Baltimore and Susquehanna Railroad at Bolton Station, Maryland. It was a single-track bridge with a clear span of 50 feet and a girder depth of 6 feet. The girder was a curious type that was much more complex and expensive than the simple flanged girder built up of riveted steel that is universal among contemporary railroad bridges. Each girder of Millholland's span was a double member made up of wrought-iron boiler plates, the two parallel sets of plates bolted together in a primitive flangeless box girder. Millholland, who was a locomotive builder, derived the double girder from the locomotive firebox, which consists of a double-walled shell rigidly interconnected by staybolts. It was a crude form, but it served well enough to warrant its replacement with a double-track structure of the same kind, which survived until 1882, when the single-web flanged girder became the standard form.

The invention that was ultimately to play the greatest role in the construction of truss bridges was the truss patented by Caleb and Thomas Pratt in 1844. Thomas Pratt's training as well as his invention indicates that a new chapter had begun in structural history, one in which science was moving increasingly to the forefront. His father, Caleb, was a Boston architect who implanted the interest that led Thomas to study building construction, mathematics, and natural science at the Rensselaer Polytechnic Institute at Troy, New York, before he entered into a long and fruitful career that began with the Army Engineers and reached its most productive years on the engineering staffs of a number of New England railroads. The Pratt truss of 1844 was designed for construction in iron alone or in iron and timber, with flat parallel chords or with an arched top chord (Fig. 31). In appearance the various forms differed little from the standard Howe truss, but the action of the web members was exactly reversed. The diagonal members were in tension and were accordingly wrought-iron rods, and the vertical members were in compression and hence acted as true posts, to be built either of wood or cast iron. The superiority to the Howe truss lay in the restriction of the longer diagonals, which were most likely to buckle, to tension. Of the two diagonals in each panel, one, the counterdiagonal, would ordinarily be unstressed, although it might be subject to tension under moving and eccentric loads. The arched top chord was dictated by the need for maximum depth at the center of the truss, where the bending moment reaches a maximum. Before Pratt's death in 1875 it was recognized that the double diagonals were redundant, making the truss an indeterminate structure, and so the diagonals were reduced to one in each of the end panels, then to one in all but the two center panels, and finally to a single-diagonal system throughout the length of the truss.

Iron Pratt trusses were rare before the end of the Civil War, but the composite forms soon became common on the railroads whose engineering work was under Thomas Pratt's charge. As in the case of other trusses we have described, hesitant bridge builders often added supplementary arches to the spans built with iron Pratt trusses. A good example was the six-span double-track bridge built by the Pennsylvania Railroad about 1860 at Coatesville, Pennsylvania. The most spectacular of the pure truss form is the bridge built by the New York and Erie Railroad over the Genesee River at Portage, New York, to replace the wooden structure that burned in 1875. The chief engineer of this masterpiece was George S. Morison, one of the leading engineers of long-span bridges in the last quarter of the century. In the Portage bridge Pratt deck trusses of cast and wrought iron varying from 100 to 116 feet in length alternate with 50-foot spans that lie over the high braced bents of the supporting structure.

By the time Morison designed the Portage span, the method of calculating the magnitude and character of the forces in the members of a bridge truss had been placed on a scientific basis and worked out to a reliable degree of mathematical exactitude. This was the achievement of theorists, beginning with Emiland Gauthey, who had been working at the problem since the turn of the century. Foremost among them in the United States was the engineer and inventor Squire Whipple (see pp. 99–100). The practice at the time of the Portage bridge was to calculate the size of individual members in the truss on the basis of a maximum working stress in tension or compression, a stress which was always a rather small fraction of the known ultimate strength of the iron. The ratio of this strength to the working stress is known as the factor of safety. It has been constantly reduced as the quality of the metal has improved and become more predictable and as the science of structural design has achieved progressively greater exactitude. In the case of the Portage bridge Morison carefully calculated the size of every member for traffic and wind loads and then supervised a work force that erected the span in eighty-two days during the spring and summer of 1875. He and his men did their work well: the bridge still carries the traffic of the Erie-Lackawanna Railroad's secondary Buffalo line.

Subdivided forms of the Pratt truss came into prominence for long-span bridges after 1885 (see pp. 143–44). Up to the last decade of the century, however, another type was to enjoy the greatest popularity for railroad bridges. Its inventor, Squire Whipple, is the most celebrated American inventor of bridge trusses. Born at Hardwick, Massachusetts, in 1804, graduate of Union College, farmer, school teacher, railroad and canal surveyor, Whipple was thirty-seven years old before he embarked on the career that soon brought him international attention. His first patent was granted in 1841 for an iron bowstring truss that was like the original Pratt truss except for an arched top chord that was carried down to the bottom chord at the ends. Although this is a valid truss form, Whipple thought of it as an iron arch with a trussed stiffening system, a fact that is revealed by the inclined skewbacks in the abutment designed to take the tangential thrust of the arch. As a truss the bowstring survived for short highway spans until well after 1900 and is still used in wood to support vaulted roofs.

The truss which bears Whipple's name was first built in 1846 and patented the following year (Fig. 50). The unique feature was the presence of diagonals that crossed two panels rather than one and sloped downward in opposite directions from end to center. All diagonal web members and the bottom chord were designed for tension, with the posts and top chord in compression. By 1860 Whipple's design seemed to offer the best solution to the requirements of the railroad truss: it was adequately braced in the horizontal plane, the stresses in the various members were properly distributed, and it possessed the greatest over-all strength—virtues that led to its being widely adopted for long-span bridges in the two decades from 1865 to 1885.

An early and typical example of a rail span with Whipple trusses was the bridge built between 1852 and 1853 by the Rensselaer and Saratoga Railroad near Troy, New York. Whipple calculated the depth of the truss and the size and proportions of the individual members for a live load of 2,000 pounds per lineal foot of deck on a single-track span with a 146-foot clear length. The top chord and posts were hollow cast-iron cylinders, the diagonals wrought-iron rods, and the bottom chord a succession of wrought-iron eye bars. All members were connected by pins at the joints, and the two parallel trusses were joined and braced by means of transverse beams and double-diagonal wrought-iron rods in the top and bottom frames. Complete lateral bracing would have required bracing in the portal frames at the ends of the trusses, but this is the only feature of mature bridge design missing in the Troy span. The railroad replaced the bridge in 1883 because the thirty-five-ton locomotive weight common when it was built had doubled in the intervening thirty years.

In the long run perhaps Whipple's greatest contribution to the technology of bridge construction was his own treatise on the subject, originally published in 1847 under the title, *A Work on Bridge Building*. A classic of structural science, it was the first comprehensive American work on the design of truss bridges and on the physical properties of wood and iron. Whipple worked out an accurate method for calculating the nature and magnitude of the forces acting in the individual members of the truss. Knowing the strength of the materials, he could readily ascertain the allowable unit stress in tension or compression and could then calculate the size of the individual member and the depth of the truss for a given load. His calculations were always based on the most unfavorable position of the load, which would ordinarily be at the center of the truss, and on a working stress equal to one-sixth to one-fourth of the known strength of the metal. What now strikes us as a curious feature of Whipple's book is his substitution of rhetorical exposition for the trigonometric functions and algebraic equations of stress analysis. He did this deliberately in order that the reader trained only in arithmetic could follow his method, a necessity that indicated the low educational level of men then engaged in the building of bridges.

The remainder of Whipple's treatise is a handbook of bridge construction, including tables of the tensile and compressive strength of wood and iron. He is careful to point out, however, that because of lack of uniformity in the metal, loss of strength from fatigue, and other incalculable elements requiring a generous safety factor, the allowable stresses in actual practice should never exceed one-quarter of the figures he gives. The work is an admirable example of the union of theoretical and practical knowledge; yet in spite of Whipple's reputation as a builder, the first edition of the book was little read by practicing engineers. When Hermann Haupt published his *General Theory of Bridge Construction* in 1851, he was under the impression that he had written the pioneer American work on the subject. He in fact independently developed many of the ideas in Whipple's book, but his presentation of them is much less satisfactory.

Whipple demonstrated his inventive faculty once again in 1849 when he built the first Warren-truss bridge in the United States without knowledge of its English precedent. The truss was first invented in England by James Warren and Willoughby Monzani, who were granted a patent on it in 1848. It was the simplest of all the forms developed up to that time: in the original design there were no posts, and the diagonals alternately sloped in opposite directions (Fig. 32). In later modifications, posts and a second set of diagonals were added. Since Warren thought that tension and compression would be distributed in alternate diagonals, many early Warren trusses were designed with cast- and wrought-iron diagonals in alternate succession. It was subsequently learned, however, that any one diagonal may be subject to either stress or to both in succession as the load moves over the bridge. The Warren truss was not popular in the United States until near the end of the nineteenth century, and then only in the form with posts and single diagonals. It and the single-diagonal Pratt truss are actually the most efficient forms, with the consequence that they rapidly superseded all other types after 1900.

The construction of the Baltimore and Ohio Railroad through the Allegheny Mountains to the Ohio River at Wheeling called forth the inventive talents of a brilliant staff of engineers under the direction of Benjamin Latrobe II. First among them was Wendel Bollman, who was granted a patent in 1852 for what he called his "suspension truss bridge" (Fig. 33). This odd-looking form was essentially a Pratt truss on which a system of wrought-iron rods radiating downward and outward from the upper corners of the truss was superimposed. These rods functioned very much like the hangers of a suspension bridge in that they carried a high proportion of the tension normally placed on the bottom chord. The form was highly redundant unless the diagonals of the truss panels were omitted, in which case the basic structure ceased to be a rigid truss. The questionable principle behind it is similar to the one we have noted in the case of trusses with supplementary arches, namely, that the strength of a bridge can be increased by adding together two different types of bridge structure.

The Bollman truss (the original example was built at Laurel, Maryland, in 1850) was popular for about twenty years, especially on the B. and O. lines east of the Ohio River. The seven-span bridge built between 1862 and 1868 at Harpers Ferry had one curious feature that may have been unique. The two trusses of the span near the Maryland bank were splayed out to make room for the switch and turnout of a branch that joined the main line at that point (Fig. 33). Between the bank and the widened span the two lines were carried by separate spans where the tracks curved sharply away from each other. The railroad replaced the Harpers Ferry bridge in 1894, but it survived as a highway span until it was destroyed by flood in 1935. The largest of all bridges incorporating Bollman trusses was the railroad's first Ohio River crossing, built from 1868 to 1871 between Bellaire, Ohio, and Benwood, West Virginia. The iron structure, exclusive of approaches, was 3,916 feet in over-all length; of the fourteen spans over water, nine were carried by Bollman trusses, with a maximum length of 125 feet, and the remainder by Whipple trusses.

The second of the B. and O.'s engineers to achieve fame in bridge design was Albert Fink, who was born and educated in Germany and who eventually became a vice-president of the Louisville and Nashville Railroad. The truss which he patented in 1854 was another case of a highly redundant design, marked by a multiplicity of diagonal members (Fig. 34). In addition to the double-diagonal panels of the Pratt truss, Fink added single diagonals crossing two panels and radiating bars extending from the upper corner of the truss to alternate panel points along the line of the bottom chord. It was indeterminate, inefficient, and redundant, but there was no question of its strength, and it served railroad needs well enough during the two decades of its active life. Smaller spans of Fink trusses were usually built with chords and posts of timber, but the larger were entirely cast and wrought iron. A curious feature of the Fink deck truss (one which lies below the deck of the bridge) was the omission of the bottom chord, the tension being taken entirely by the diagonal members.

The first iron structure to embody Fink's design was the B. and O. bridge over the Monongahela River at Fairmount, West Virginia, built between 1851 and 1852 (Fig. 34). His most spectacular work was his first major undertaking for the Louisville and Nashville, the deck-truss bridge constructed between 1857 and 1859 to cross the Green River near Mammoth Cave, Kentucky. Contemporary prints of this high single-track structure convey a precarious spider-web quality that arouses nervous apprehension today. According to the standards of rail loading that came to prevail at the time of World War I, it was flimsy indeed; yet the thin, flexible bars and rods that composed its trusses were perfectly stable for its day. Fink rounded out his career as a bridge builder with another innovation in truss design when he was chief engineer for the first of the long bridges to cross the Ohio River at Louisville, Kentucky, a work that we will consider under long-span bridges (see pp. 142–43).

The decade of the 1840's, which produced the great basic inventions in truss form, might be regarded as the heroic age of the iron bridge. The following decade saw a steady stream of patents granted to inventors who followed in the wake of the pioneers. The construction of twenty-one thousand miles of railroad line between 1850 and 1860 again acted as a powerful stimulus to this creative urge. It was a prolific age, but most of the inventors were distinguished more by zeal and ambition than by sound scientific knowledge. Some of the innovations were modifications of existing forms, a few of which proved useful and were eventually adopted as standard features of bridge design. Of the many new truss forms patented in the fifties, the majority were fantastic: either they had no structural validity, or they were so ill-conceived as to be prohibitively expensive or impossible to analyze. Among new types the lenticular or lense-shaped truss had the merit of providing maximum depth at the center, but there was little point to curving the bottom chord as well as the top. The largest lenticular trusses were incorporated in the Smithfield Street Bridge over the Monongahela River at Pittsburgh (1882–84), the work of Gustav Lindenthal. The lattice truss was adapted to iron in 1859 and was shortly afterward used for the Allegheny River crossing of the Pittsburgh, Fort Wayne and Chicago Railroad at Pittsburgh. The form continued in favor until about 1910, when a number were built by the Chicago and North Western Railway for its Chicago-Milwaukee freight line. The tied arch of iron was another product of the fifties: in this type the horizontal thrust is taken by a series of ties extending between the springing points rather than by piers or abutments. The tied arch is used wherever conditions are such that abutments of sufficient depth cannot be built.

The last truss to come in the period of basic developments was invented by S. S. Post, an engineer with the New York and Erie Railroad, who was granted a patent in 1865. The Post truss was like the Pratt except that the posts were inclined, sloping inward in opposite directions on either side of the center panel. Post thought that this feature offered greater rigidity, but there is no ground for supposing that this would be the case. The largest bridge with Post trusses was the eleven-span structure built by the Union Pacific Railroad, from 1868 to 1872, to cross the Missouri River between Omaha, Nebraska, and Council Bluffs, Iowa (Fig. 35). This remarkable exhibition of daring and ineptitude was as noteworthy for its short life as for its great size and unusual construction. Each pier of the bridge was made up of a pair of cast-iron drums filled with concrete, a crude but useful innovation that constituted an early step in the development of the concrete bent. On the other hand, the Omaha span had no horizontal bracing in the top, bottom, and portal frames, so that the two trusses of any one span were virtually independent structures. This oversight made the bridge radically vulnerable to wind, which soon demonstrated the fact: most of the river spans were destroyed by a tornado in 1877. The destruction of the Omaha span did not help the popularity of the Post truss, but it enjoyed a fairly vigorous life up to the mid-eighties.

The heroic age of the truss bridge also saw the creation of the suspension bridge essentially as we know it today. The structural principle underlying such bridges is a relatively simple concept that leads to a highly efficient form. Except for the towers, the structure of a suspension bridge works entirely in tension and is thus the exact opposite of the arch. The deck is hung by vertical suspenders from two or more cables the tension in which is transmitted to the foundations through compression in the towers. Because of the flexibility of the cables and the suspenders, the deck must be stiffened by trusses sufficiently to prevent vibrations from traffic and wind that would otherwise render the structure useless.

Bridges with walkways suspended from bamboo rope were built in China as early as the first century B. C., and iron chains were substituted for bamboo as early as the sixth century. Yet the suspension bridge remained unknown in Europe until the sixteenth century, and even then it existed only as an idea. It was not to appear in actuality until 1741, when a footbridge hung from iron chains was built over the River Lees in Durham County, England. This precedent was not followed in Europe until the nineteenth century, with the result that the United States was able to establish leadership in this type, as it did in the case of the timber truss. The pioneer was James Finley, a judge of Fayette County, Pennsylvania, who invented a flexible chain of wrought-iron links in 1796. He built his first bridge in 1801 to span Jacobs Creek in Uniontown, Pennsylvania. Both the cables and the suspenders were wrought-iron chains, and the deck was maintained rigidly in a level position by longitudinal stiffening trusses of wood. The chain, held in place by masonry anchors buried behind the abutments, passed over wood-block saddles in the timber-framed towers. It was the prototype of all subsequent suspension bridges, the evolution of which required surprisingly few modifications in the fundamental elements.

Finley was granted a patent for the level-deck iron-chain bridge in 1808. The method he gave for laying off the parabolic curve of the cable offers an example of how homely ingenuity may serve where mathematics is wanting. A flexible cord carrying only its own weight will fall naturally into a catenary, but if the cord is loaded uniformly on the horizontal line, it will assume a curve between a catenary and a parabola. With analytic geometry Finley could readily have calculated the shape of this curve for a given span and rise. His method, however, was to load a string by attaching to it weights equally spaced on the horizontal line, the points of attachment representing the transverse beams of the bridge deck. Having already determined the length of the span and the tower height, he then traced the curve on a board directly from his loaded string. The design of a great suspension bridge like George Washington or Golden Gate would be a vastly more complex matter, but the determination of the cable curvature remains one of the essential problems.

Finley built a number of bridges in New England and in the Delaware and Potomac valleys. One over the Potomac above Georgetown, completed in 1807, gave the area the still-current name of Chain Bridge. The little 100-foot span came to play an important role in the evolution of the form. Albert Gallatin included a description of it in his *Report on the Roads and Canals of the United States* (1808), as did Thomas Pope in his *Treatise on Bridge Architecture* (1811). Both these works were known to Louis Marie Navier, who included a discussion of the structure in his *Rapport et Mémoire sur les Ponts Suspendus* (1823), the great classic on suspension bridge design. The French authority gave Finley credit for inventing the rigid level-deck bridge and thus provided the avenue through which his work came to be known among the European builders. There is also the possibility that Finley's bridges were known to Thomas Telford, who may have been influenced by them in his design of the celebrated suspension bridge over Menai Strait in Wales (1818–26), the largest chain span in existence at the time.

The longest span among Finley's bridges was the Schuylkill River bridge at Philadelphia, which opened in 1809. Limited to pedestrian traffic, the narrow walkway extended 308 feet between the towers. The bridge collapsed under a load of snow and ice in 1816 and was replaced the same year by a bridge with a span between towers of 408 feet that was suspended from wire cables. Built by Josiah White and Erskine Hazard, owners of a Philadelphia wire mill, it was the first wire-cable bridge in the United States. Wire cable offered two great advantages as opposed to chains: first, drawing wire through dies is a simpler and less expensive process than the manufacture of hand-wrought iron links; second, the tensile strength of wire is extremely high because the process of drawing the wire down to successively smaller diameters increases its strength with each reduction in diameter. If a large number of wires are bound into a cable, the resulting member will have a tensile strength several times higher than that of a solid member with an equal cross-sectional area of metal.

White and Hazard's footbridge at Philadelphia established the feasibility of the wire-cable span, but the first builder to see its immense possibilities was Charles Ellet. Born in Bristol, Pennsylvania, in 1810, a self-taught surveyor for the Chesapeake and Ohio Canal, he went to Paris in 1830 and enrolled for one year at the École Polytechnique. He was the only American engineer of the time who studied at this celebrated scientific and technical school. His enthusiasm fired by the work of the theorists and inventors who were creating the European suspension bridge, he returned to the United States full of grand schemes for wire-cable bridges of unprecedented size. Few of his projects were realized, and among these one suffered partial collapse and a second was brought to completion by another engineer. In 1832 Ellet proposed a bridge over the Potomac River at Washington with a span of 1,000 feet; seven years later he proposed an even grander one over the Mississippi at St. Louis with a 1,200-foot span. His 1848 project for a bridge over the Connecticut River at Middletown, Connecticut, would have been the first railroad suspension bridge, but it was opposed by the nearly unanimous opinion of the engineering profession, whose members held that bridges constructed of wire would lack the strength and rigidity necessary to support moving trains.

Ellet's first completed structure was built over the Schuylkill at Philadelphia between 1841 and 1842 (Fig. 36). The 358-foot main span was carried by wire-rope suspenders hung from ten unbound wrought-iron cables, five on each side. The slightly bowed deck was stiffened by low Howe trusses of wood that also acted as hand rails. The timber frames of the towers were covered with wood sheathing. The Schuylkill bridge was long ago replaced, but a similiar structure, completed in the same year at New Portland, Maine, still carries a light highway traffic. Ellet's great work was the Ohio River bridge at Wheeling, West Virginia, constructed between 1846 and 1849, and its fate was one of the many reversals of fortune that dogged the engineer's whole life. The main span of the bridge extended 1,010 feet between towers, which were the first built of masonry in an American suspension bridge. The wire-rope suspenders hung from twelve main cables, six to a side, each containing 550 parallel wrought-iron wires. It was a heroic undertaking for its day, but its instability in high winds led to its partial destruction by a storm in 1854. It was rebuilt the same year by John Roebling, who introduced the wire-rope stays radiating downward from the tops of the towers to the lower ends of the suspenders in order to give the bridge aerodynamic stability. Roebling was far ahead of his time in his intuitive grasp of the aerodynamic problem, as the collapse in 1940 of the Tacoma Narrows bridge showed (see p. 238).

Ellet's most difficult project was the Niagara Falls bridge, the auxiliary span for which he built between 1847 and 1848. The need for a bridge in this location and the problems fixed by the site offered an extraordinary challenge to the engineer. The gorge of the Niagara River has nearly vertical sides above the talus slopes at the foot of the cliffs and varies in width throughout most of its length from

800 to 1,000 feet. The number of rail lines terminating on both sides of the Niagara frontier, even at that early date, clearly established the economic value of a bridge near the Falls, in spite of the formidable difficulties in the way of constructing one. The depth of the gorge and the swift, turbulent current of the Niagara River made it necessary to construct a bridge as a single span. Until the steel arch or the cantilever truss was sufficiently developed to cross the stream in one leap, the suspension bridge offered the only feasible alternative. The structure which Ellet completed in 1848 was a light vehicular and pedestrian bridge with a 770-foot span and a 9-foot width. His intention was to use it as an auxiliary construction span for a much heavier rail-highway bridge that was to be the permanent installation. But Ellet became involved in a dispute with the directors of the bridge company that led to his resignation as chief engineer in 1848. Three years later John Roebling was appointed to replace him and ultimately carried the project to its successful completion. This unfortunate controversy marked the end of Ellet's career as a bridge builder.

The man who followed him on the Niagara project had already established himself as the leading authority on the construction of the suspension bridge; before his career was ended by death two decades later, he was to prove himself as the most courageous and imaginative builder of his time. John August Roebling was born in Mühlhausen, Germany, in 1806, graduated from the Royal Polytechnic School of Berlin twenty years later, and emigrated to the United States in 1831 because his republican convictions blocked his career as a civil engineer under the Prussian government. After an abortive attempt at farming in eastern Pennsylvania, he returned to engineering as a surveyor for the Beaver River Canal and later the Pennsylvania Railroad. He developed a method for manufacturing wire cable, and founded a company at Trenton, New Jersey, in 1849 to exploit his invention. Roebling had been interested in the suspension bridge since he had been a student, having written a thesis on the subject as part of his graduation requirement, but he was thirty-eight years old before he won his first commission in 1844. For the next 25 years, to his death in 1869, he enjoyed the most productive bridge-building career of his century.

The first five of Roebling's suspension bridges, constructed between 1844 and 1850, were all aqueducts, the first to carry the Pennsylvania State Canal over the Allegheny River, the others to carry the Delaware and Hudson Canal over various streams in northeastern Pennsylvania. The complex of commercial and technical determinants which shaped these curious but impressive structures belonged to a phase of economic history that was already beginning to pass. The once extensive canal system of Pennsylvania was rapidly rendered obsolete by the expansion of the railroads in the mid-century decade. In the short period in which these waterways flourished, Roebling was repeatedly able to provide demonstrations of the validity of the suspension principle for supporting extremely heavy loads. The structural system was uniform for all the aqueducts: they were multispan suspension bridges in which a continuous troughlike flume of planking was fitted into a timber frame of transverse floor beams and side-wall trusses; this frame was suspended between wrought-iron cables extending over squat stone posts that were carried on masonry piers. It was massive construction that was perfectly adapted to carrying the great weight of water but one that was limited entirely to this peculiar requirement. Even the multispan feature has disappeared because of the flexibility of the suspension bridge.

When Roebling took over the Niagara project following Ellet's resignation, he had good reason for confidence in the idea of a railroad suspension bridge, although his predecessor was apparently the only engineer who shared his views. The Roebling bridge, built between 1851 and 1855, was a double-deck structure with an 821-foot span extending between masonry towers. The lower deck was a roadway and the upper the joint single-track line of the Grand Trunk and the Great Western Railway of Canada. The chief supporting members were four 10-inch cables, two to a side, one for the upper deck, the other for the lower. Each cable contained 3,640 wires, laid parallel and tightly bound into a functioning unit. The wire was manufactured in England because Roebling doubted whether any Canadian or American enterprise, including his own, was capable of the job. In addition to the vertical suspenders, there were radiating stays to provide aerodynamic stability and to increase the rigidity of the whole structure. The deep stiffening trusses between the decks were Pratt trusses with timber posts and wrought-iron diagonals. This classic bridge performed very well for forty years, but the increasing weight of rail traffic eventually forced its replacement in 1896 by the steel arch bridge that still serves the Canadian National Railway and its New York connections. It was the beginning and the end of the railroad suspension bridge: it proved to be too light and flexible for the loads and speeds of twentieth-century rail traffic. The community at the east end of the Roebling span, although now incorporated in the city of Niagara Falls, is still called Suspension Bridge after this once world-famous structure.

While the Niagara bridge was under construction, Roebling completed the rebuilding of Ellet's Wheeling bridge. Two years after it was opened to service, Roebling was engaged on the greatest project he was to see to completion before his death. Initial plans for the Ohio River suspension bridge at Cincinnati, Ohio, were prepared in 1856, but eleven years were to pass before the structure was completed in all its details (Fig. 37). There were compelling reasons for a bridge at Cincinnati. The community had flourished economically and culturally: by the time of the Civil War it was the fourth city in population, the leading meat-packing center of the nation, and a major focus of the Ohio River trade. Its location midway along the length of the river, equally distant from the Pocohontas coal fields and the agricultural areas of the lower Ohio valley, made it one of the main gateways to the South as well as an industrial and food-processing center. By the mid-century, the need for an Ohio River crossing was obvious, and Cincinnati was the natural location for it. But the 1,500-foot river with its disastrous annual floods and its heavy water-borne commerce presented a serious challenge to the builder: a multispan timber structure seemed hazardous, but at the same time a single-span suspension bridge seemed a daring and costly novelty. Yet plans to bridge the river between Covington, Kentucky, and the larger city were first proposed in 1845 and discussed with growing seriousness during the next ten years. The citizens of both communities finally made the decision for a suspension bridge, engaged Roebling as chief engineer, and accumulated enough capital to begin construction in 1856.

The natural obstacles offered difficulties enough, but these were compounded by war and financial failure. Only a man with Roebling's courage, patience, and confidence could have seen it through. His first problem was to find suitable bearing material in a region where river sediments overlie thin limestone strata interspersed with shale beds little harder than ordinary clay. Roebling found a thick bed of well-compacted gravel not far below the surface of the alluvial material, excavated the area of the tower footings on either shore down to the gravel, and began to lay foundation courses before the end of the year. It was then that troubles began to multiply: ice and floods in the winter and spring of 1856–57 stopped construction until early summer; the panic and depression of 1858 threatened the financial soundness of the company and caused a second stoppage; then the Civil War intervened and prevented work entirely from 1861 to 1863. When construction was resumed the following year, storms three times destroyed the auxiliary footbridges erected to string the cable wires. Roebling's tenacity carried it to a successful conclusion, and the bridge was opened to traffic in an unfinished state on New Year's Day, 1867, and finally completed during the summer of that year. The finished work set two records, one for its main-span length of 1,057 feet, the other for the 230-foot height of the masonry towers. The familiar radiating stays provide stability against wind, supplemented by wrought-iron stiffening trusses in the Pratt form. This ironwork was extensively reconstructed about 1898 to provide greater strength to carry streetcar traffic. The modified bridge still stands in full use, although the streetcars have been replaced by buses.

During the early years of construction on the Cincinnati span, Roebling was engaged on a smaller project at Pittsburgh. The Sixth Street bridge over the Allegheny River, built between 1857 and 1860, was noteworthy for two reasons. For the first time John Roebling's son, Washington, was involved in the design and construction of the work, acting as his father's assistant in the drafting room and on the site. More important for structural history, construction of the bridge was accomplished by spinning the cable between anchors and over the towers by means of traveling sheaves. The looped ends of the cable strands were passed through the eyes of wrought-iron eye bars, which were in turn fixed to cast-iron plates buried in the masonry anchor piers. The previous method of installing cable was to prefabricate the strands on the ground and raise them into position for binding, by means of barges, falsework, and floating cranes. Without the method of cable-spinning developed by Roebling, a bridge with the magnitude of the Cincinnati span would have been an impossible undertaking.

The greatest of the Roebling projects is Brooklyn Bridge, but the old master's death at the beginning of construction meant that it was Washington's achievement as much as his father's. Moreover, Brooklyn Bridge belongs more fully than the previous works to the new age of scientifically designed long-span bridges that was to come with the great period of industrial and railroad expansion after the Civil War. We will treat these structures in their appropriate chapter (for Brooklyn Bridge, see pp. 152–54).

Part Three

The Rise of the Industrial Republic, 1865–1900

Chapter Nine | The New York Skyscraper

The peculiarly American phenomenon known as the skyscraper was a necessary consequence of three factors that emerged in full force after 1865: economic expansion, the financial and institutional character of American business, and the intensive use of land in the main commercial centers. The skyscraper was the most conspicuous result of these changing patterns, which left their mark on the whole city. In the early decades of the republic ecclesiastical and civic buildings dominated the skyline; after the Civil War office buildings, hotels and apartments, factories, and railroad stations rapidly eclipsed all other structures in size and number. New utilitarian demands and new structural materials and techniques radically transformed the large urban buildings. Iron gave way to steel; bearing walls were replaced by curtains; masonry fell before concrete; and the craft technique of building was progressively transformed into a scientific technology. The beginnings of these processes occurred before the Civil War; the drastic and often violent consequences came after it.

The question whether the skyscraper originated in New York or Chicago is still a matter of controversy, for the answer depends on whether we define it in terms of size and economic function or in terms of structural and architectural character. If we think of the skyscraper as a high commercial building whose height greatly exceeds its horizontal dimensions and which grew out of economic exigencies arising from intensive land use, then the form may be regarded as the creation of New York builders. On the other hand, if we hold that the structural system of wind-braced steel or concrete framing expressed in an organic architectural form is an essential characteristic, then the skyscraper was a Chicago achievement. Indeed, the structural techniques used by the New York builders were for many years thoroughly conservative and lagged well behind the level reached in Chicago. It is ironic that the appearance of the New York skyscraper coincided with a return to the masonry bearing wall, which came about simply because the builders failed to exploit the potentialities of a cast-iron front. The idea that the origin of the skyscraper is as much bound up with structural innovation as with great height is suggested by the earliest uses of the term. The word appeared almost simultaneously in two widely separated sources, but

42. Home Insurance Building, Chicago, Ill., 1884–85. William Le Baron Jenney, architect. This celebrated building came closest to being the first true skyscraper, with fully developed skeletal construction.

43. Metropolitan Building, Minneapolis, Minn., 1888–90. E. Townsend Mix, architect. Interior of the light court.

44. Fair (Montgomery Ward) Store, Chicago, Ill., 1890–91. William Le Baron Jenney, architect. Part of the steel and wrought-iron frame during construction.

45. Unity Building, Chicago, Ill., 1891–92. Clinton J. Warren, architect. The developed iron and steel frame for a Chicago skyscraper.

46. Garrick Theater Building, Chicago, Ill., 1891–92. Adler and Sullivan, architects. The trusses supporting the office floors over the theater, exposed during demolition in spring, 1961.

47. Grand Central Terminal, New York City, 1869–71. John B. Snook, architect; Isaac C. Buckhout, engineer. Interior of the train shed. The track area of the original Grand Central was the first in the United States to be covered by the great vaulted roof later called the balloon shed.

48. Reading Terminal, Philadelphia, Pa., 1891–93. F. H. Kimball, architect; Joseph M. Wilson and Brothers, engineers. The rear end of the train shed. The photograph clearly shows the hinged arches supporting the roof.

49. Lackawanna Terminal, Hoboken, N. J., 1904–06. Kenneth W. Murchison, architect; Lincoln Bush and E. W. Stern, engineers. Rear end of the train shed. The roof of low parallel vaults, invented by and named after Lincoln Bush, was popular up to about 1920.

48

49

50. Bridge of the Pittsburgh and Steubenville Railroad, Ohio River, Steubenville, Ohio, 1863–65. Jacob H. Linville, engineer. Whipple truss of the channel span. Regarded as the first of the long-span railroad bridges, the Steubenville bridge was built by a company that later became part of the Pennsylvania system.

51. Bridge of the Chesapeake and Ohio Railway, Ohio River, Cincinnati, Ohio, 1886–88. William H. Burr, engineer. Subdivided Pratt truss of the channel span.

52. Bridge of the Kansas City, Fort Scott and Memphis Railroad, Mississippi River, Memphis, Tenn., 1888–92. George S. Morison and Alfred

50

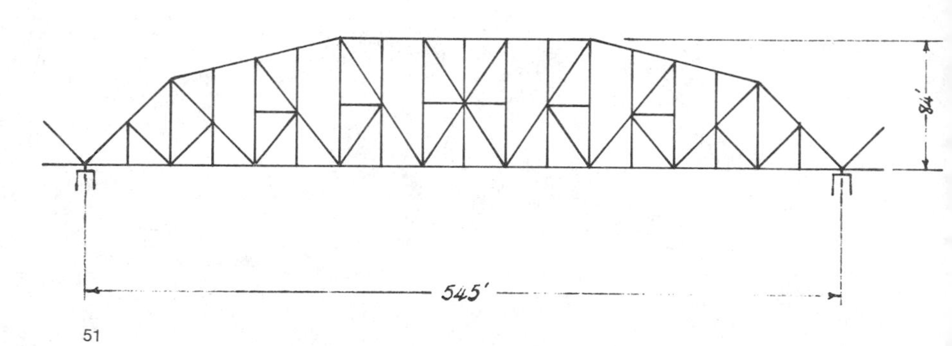

51

52

Noble, engineers. The Memphis bridge was typical of the long-span rail cantilevers designed according to the most advanced scientific principles of the day. The company that built it was later merged with the St. Louis–San Francisco Railway.

53. Bridge of the Galveston, Harrisburg and San Antonio Railway, Pecos River, Comstock, Tex., 1891–92. Extraordinary height rather than long span made this cantilever structure a spectacular work of bridge engineering. The building company is now the western portion of the Texas and New Orleans Railroad, a Southern Pacific subsidiary.

53

54. Eads Bridge, Mississippi River, St. Louis, Mo., 1868–74. James B. Eads, chief engineer. The arch bridge in the United States suddenly came of age with Eads's magnificent work, the first American structure to be built of steel.

55. Washington Bridge, Harlem River, New York City, 1887–89. William R. Hutton, engineer; Edward H. Kendall, architect. All the essential features of the modern steel arch bridge are present in the Washington span.

54

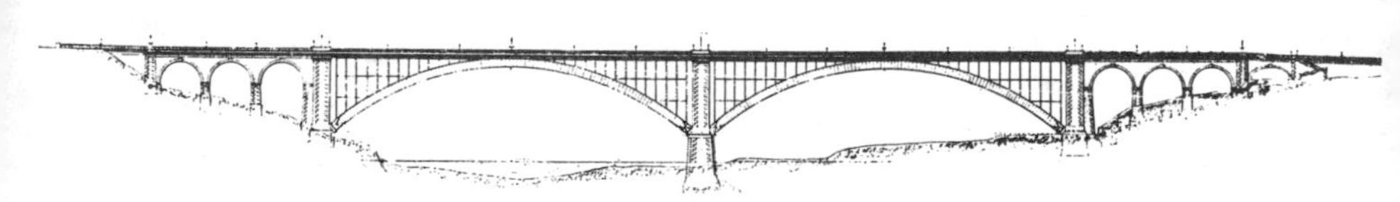

55

56. Brooklyn Bridge, East River, New York City, 1869–83. John and Washington Roebling, engineers. The greatest of the Roebling suspension bridges, Brooklyn is so well known as to be a national monument.

57. Hotel Ponce de Leon, St. Augustine, Fla., 1886–88. Carrère and Hastings, architects. The hotel was the first multistory public building in the United States to be constructed of concrete.

58. South Fork Dam, South Fork of Conemaugh River, near Johnstown, Pa., 1839. William E. Morris, engineer. Cross section of the dam and details of the control works. South Fork was a mature example of the earth- and rock-fill dam; careless maintenance led to the collapse that produced the Johnstown flood in 1889.

59. Sweetwater Dam, Sweetwater River, near San Diego, Calif., 1886–88. James D. Schuyler, engineer. Cross section through dam, gate house, and sluiceway. The original portion of the dam was the first to be constructed entirely of concrete.

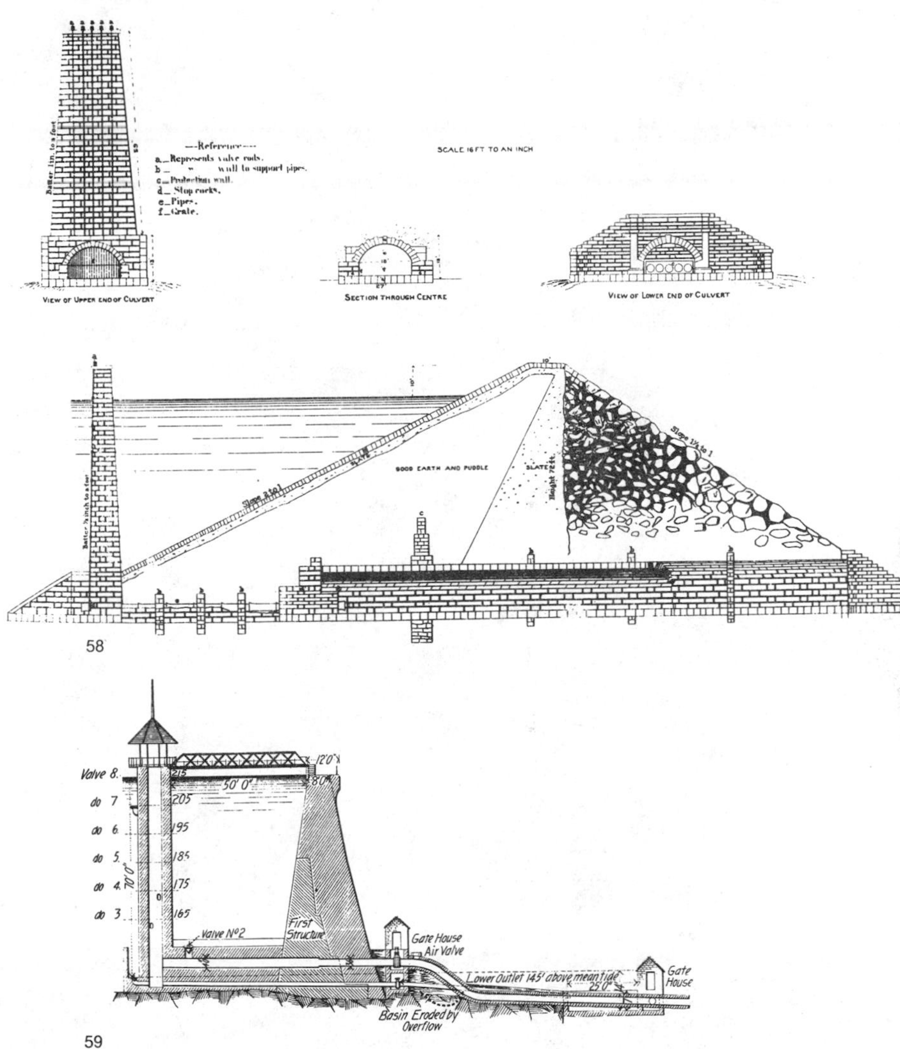

60. William E. Ward house, Port Chester, N. Y., 1871–76. Robert Mook, architect. Ward's private residence was the first building in the United States to be constructed entirely of reinforced concrete.

61. *Top*: Ribbed reinforced concrete floor slab, 1885. *Bottom*: Arched reinforced concrete floor slab of Bourn and Wise Wine Cellar, St. Helena, Calif., 1888. Ernest L. Ransome, engineer.

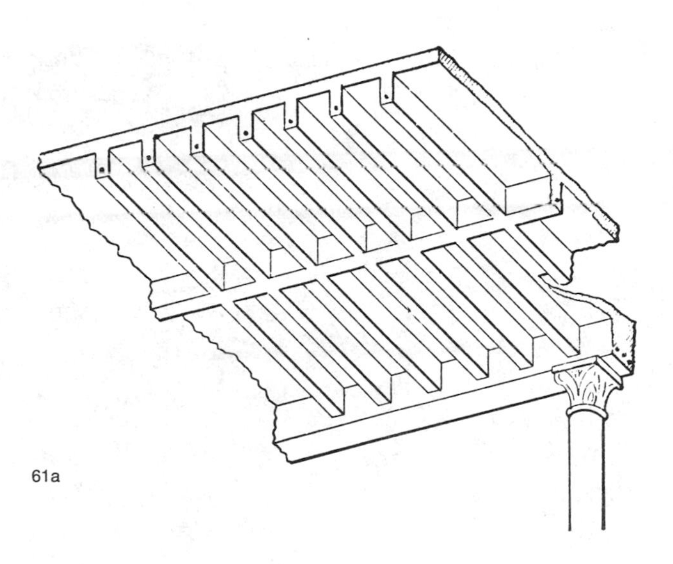

61a

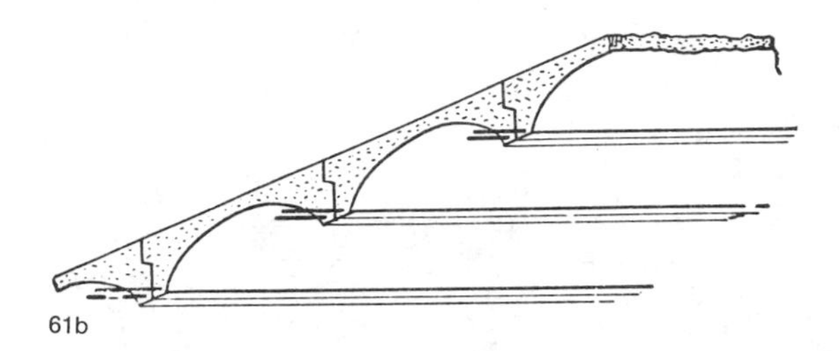

61b

62. Alvord Lake Bridge, Golden Gate Park, San Francisco, Calif., 1889. Ernest L. Ransome, engineer. The first reinforced concrete bridge in the United States.

63. Melan reinforced concrete arch, Franklin Bridge, St. Louis, Mo., 1898. Longitudinal section. The Melan system of reinforcing bridges with heavy ribs was effective but inefficient and was superseded by the more flexible bar reinforcing after the turn of the century.

62

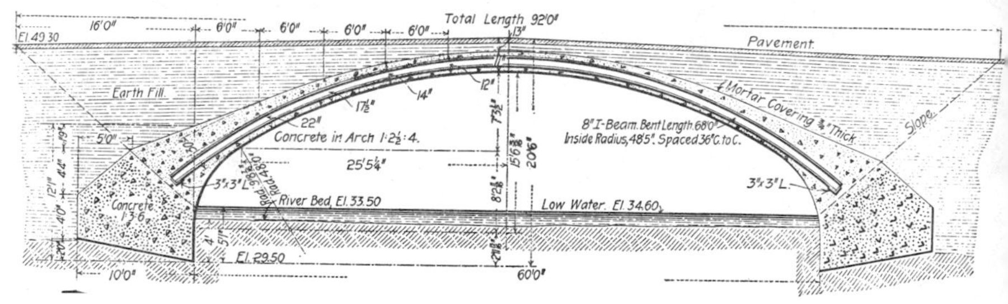

63

in both cases the context was substantially the same: it came first in an article entitled "Chicago's Skyscrapers" in the *Chicago Tribune* for January 13, 1889; second, in an article concerned mainly with Chicago building in the issue of *Engineering News* for January 19, 1889. The latter, edited and published in New York, was the leading journal of the construction industry, as its successor continues to be. The presence of the term in connection with Chicago in a national journal implies that the concept was closely associated with the technical achievements of the Chicago builders.

Yet the economic conditions that required the construction of very high buildings on narrow lots first became decisive in New York. Intensive land use, with the resulting speculative traffic in real estate, was itself a consequence of more powerful and pervasive forces. Industrial expansion was the primary factor, and in the period following the Civil War much of that expansion continued to be concentrated in New England and the Middle Atlantic states. Financial institutions had long been concentrated in Boston, but the center was rapidly shifting to New York around the time of the war. The close interdependence of the new industrial pattern demanded centralization of business administration. The leadership of New York as an ocean port, along with its proximity to major arteries of land transportation, was a local factor that helped to make the city an irresistible magnet. Its attractive power, once established, has never waned: even today it has no serious competitor as the primary administrative and banking center. The financial resources for the high building were there; the development of safe elevator transportation (chiefly by Elisha Graves Otis) and the organization of construction facilities soon made it a reality.

The first of the new towers was the office building of the Equitable Life Assurance Company, constructed at 120 Broadway between 1868 and 1870. It was originally designed by Arthur Gilman and George H. Kendall to be built entirely of masonry, but before construction began the young architect and engineer George B. Post was commissioned to redesign the building because of the high cost of the original program. The result was a tower that rose to an over-all height of 130 feet, although there were only five working stories within the main enclosure. Post kept the exterior granite piers but introduced a far from novel construction of brick partitions, brick floor arches, and wrought-iron beams in the interior. The changes greatly reduced the cost of construction of what proved to be a highly profitable investment. A success of this magnitude was sufficient to launch Post on a long and prosperous career.

The chief reason for the architect's reluctance to use iron framing throughout the interior was the justified fear that exposed metal would buckle and even collapse in fire. But several inventions were soon to overcome this danger. The first was a method of covering floor beams with a cladding of fireproof tile, patented by Balthasar Kreischer and George H. Johnson in 1871. The latter shortly extended the technique to the covering of columns and first applied both inventions to the interior iron members of the Kendall Building in Chicago (1872–73). A similar method of protecting iron columns was patented by Peter B. Wight of Chicago in 1874. Mechanical improvements in the speed and safety of elevators, the progress of iron framing, and the reliable fireproofing of ironwork—all these eventually made the Equitable Building seem old-fashioned, but the New York engineers and architects, including Post himself, were slow to take advantage of the new techniques. Economic factors doubled the height of the New York skyscraper within five years: Post's Western Union Building rose to 230 feet, and Richard M. Hunt's Tribune to 260 feet. For all their unprecedented height, however, they relied on the same combination of masonry walls and partitions supporting wrought-iron beams. It was easy to see the costly waste of space in the use of masonry elements on this scale, but no one in New York was willing to do much about it for another decade.

The decisive step was at last taken by Post in his design of the Produce Exchange, erected at the foot of Broadway between 1881 and 1884 (Fig. 38). The eight-story, 125-foot height and the extensive horizontal dimensions hardly made it a skyscraper, but the structural system was very nearly what was needed to make the modern skyscraper a possibility. The primary requirement was to support the floor, roof, and wind loads entirely on the iron skeleton, thus freeing the partitions and walls from any bearing function and reducing the latter to protective screens. Post literally came within a few inches of achieving this end. Because of the alluvial and marine sediments overlying the bearing rock at the lower end of Manhattan, the footings of the Exchange rested on piling driven into heavy clay. The floor and roof loads were supported almost entirely by a frame of cast-iron columns and wrought-iron beams and joists. The wall was in effect a curtain, since the window elements of any one bay rested on the spandrel girder. What prevented this advanced system of framing from being a complete skeletal structure was that the wall columns were set in brick piers which carried a small proportion of the peripheral loads by supporting the ends of the spandrel girder.

The central part of the second floor in the Produce Exchange was opened to a trading room with an area of 32,000 square feet. This enclosure necessitated a radical deviation from straightforward column-and-beam framing. The skylight over the center of the trading floor and the office floors superimposed on its periphery rested on a complex system of wrought-iron trusses and widely spaced columns. This method of carrying loads over large interior openings had its beginning in the timber-truss framing of wide-span roofs. From these there was a fairly straight line of development through the wood-and-iron trusses of mills and train sheds to such masterpieces of interior truss framing as the Auditorium and Garrick Theater buildings in Chicago (see pp. 122–23, 127–29). Post's design of the Produce Exchange was an early and successful solution of the problem of carrying upper stories above an open enclosure. When the Exchange was demolished about 1960, its iron structure was still in sound condition.

One of the valuable lessons for New York builders came in connection with a construction problem that had nothing to do with skyscrapers or train sheds. The Statue of Liberty, erected between 1883 and 1886, is a copper-sheathed enclosure whose height of 151 feet 5 inches made it comparable in size if not in weight to the skyscraper of its day. The sculptor was Frédéric Auguste Bartholdi, and the engineer no less than Gustave Eiffel, who was shortly to astonish the world with an exhibition of his power as a creative builder. The problem posed by the statue was less one of supporting vertical loads than of resisting the force of wind on the extensive surface area surrounding the hollow interior. Eiffel's solution involved two valuable innovations for American building. First, the diagonally braced frame inside the figure represented the most extensive system of wind bracing so far employed for any American structure other than a bridge. The second innovation was the use of steel for the posts that constitute the chief bearing members of the frame. This was the second time in the United States and the first in New York that steel was specified for a structure other than a bridge. (On the beginnings of steel construction, see pp. 124–25, 142, 149, and 154.)

Back on Manhattan Island, the building usually regarded as the first work of skeletal construction in New York was marked by both progress and retrogression. The Tower Building, designed by Bradford Gilbert and erected at 50 Broadway (1888–89), was an odd combination of the traditional and the advanced. The first seven of its eleven stories were carried wholly by an iron skeleton: wrought-iron girders extending across the width of the narrow building were supported at their ends by cast-iron columns inserted in the brick party walls, which were thus brick curtains carried floor by floor on the spandrel girders. The same characteristic marked the façade, where the curtain was a combination of vertical bands of masonry and banks of windows. Diagonal bracing across the outer bay provided rigidity against wind loads. Up to the seventh floor Gilbert's work was a novelty among Manhattan towers; above it, however, the side walls reverted to solid bearing masonry resting on heavy girders that extended over the tops of the seventh-floor columns. The architect's purpose in designing the building in this curious way was to provide an adequate amount of rentable space on the lower floors. With a conventional masonry bearing wall for eleven stories, the clear interior width would have been only 16 feet 4 inches against 21 feet 6 inches with the iron columns. This situation arose from the absurd dimensions of downtown Manhattan lots: in the case of the Tower Building the site had a 25-foot frontage for a 108-foot depth. There was widespread prejudice against the iron skeleton for narrow buildings like the Tower, a hostility that characterized the Board of Building Examiners as well as the public. The story is that Gilbert had to occupy a top-floor office himself in order to reassure prospective but uneasy tenants.

Even the limited system of iron framing in the Tower Building was slow to be adopted in New York. The solution for the Produce Exchange seemed safer for high buildings, and Post again provided a demonstration of it on a larger scale. His fifteen-story Havemeyer Building, erected between 1891 and 1892, rose to a height of 180 feet (Fig. 39). For the most part the interior and wind loads were carried on an iron frame with double-diagonal bracing in the outer bays. Again, however, the masonry walls shared some of this burden: they carried their own weight as well as a small portion of the peripheral floor loads. The cross-sectional drawing shows how the outermost columns were set immediately inside the inner wall surface, so that the floor loads were carried mainly by the columns only up to the fourth floor. The drawing also indicates how the wall thickness was reduced in steps up the height of the building as the vertical load and the overturning force of the wind progressively decrease. Properly laid masonry walls are rigid and heavy enough to resist the force of wind up to a certain height, a characteristic that the designers of such towers as the Herald and the Western Union relied on exclusively.

The Produce Exchange and the Havemeyer Building so closely approached fully framed construction that the final step was bound to come soon and with little call on the engineering imagination, especially since the Chicago builders were now attracting international attention for high steel-framed buildings with curtain walls reduced largely to glass. The braced and riveted steel skeleton established itself once and for all in New York with the construction between 1894 and 1895 of Bruce Price's American Surety Building at Broadway and Pine Street. With a height of twenty stories, rising 303 feet above grade, a masonry wall was simply out of the question. For this height a wall without supplementary columns inserted in the masonry would have had to be at least 7½ feet thick at the base. Once the riveted steel frame had been adopted, the New York builders pushed the skyscraper higher and higher. It reached thirty stories before the end of the century, passing Chicago's record for the time, and continued its upward climb to the present limit of one hundred ten stories in the twin towers of the International Trade Center (see pp. 197–98).

Chapter Ten | The Chicago Achievement: Steel-Framed Construction

The extraordinary accomplishments of Chicago in the building and planning arts indicate that the development of the city sprang from a specially potent complex of factors not exactly paralleled in the history of any other community. Perhaps the main ingredient was the intensity of its economic growth as well as the sheer magnitude of it. Compared with the eastern cities, Chicago was late in arriving at the urban stage. By the time New York had become the leading port and the financial center of America, Chicago was little more than a collection of rude houses on a most unprepossessing site at the lower end of Lake Michigan. Incorporated in 1837, the city remained relatively isolated until the mid-century decade. The lake is an ideal avenue for commerce, but in the days of sailing vessels storms and ice were formidable hazards. Yet the location offered enormous geographical advantages, which the more farsighted citizens were soon to exploit. By the end of the Civil War the picture had changed so radically that the future appeared to offer unlimited promise.

The basic inventions necessary to the mechanization of farming and the fencing of range land opened the agricultural heart of North America to intensive development. Since Chicago was the natural center of this vast region, the first necessity was transportation. The great waterway systems of the Mississippi valley and the Great Lakes were linked in 1848, when the Illinois-Michigan Canal was opened to traffic. In the same year the first rail line began to operate trains to and from the city. With the rise of agriculture came the facilities for the storage and milling of grain, the slaughtering of cattle, and the processing and shipment of meat. The manufacture of farm machinery expanded into the basic metal-fabricating and woodworking industries. Banking and the institutions of finance soon followed. Four years after the end of the Civil War Chicago was the focal point of the largest system of inland waterways in the world and the hub of a rail network that extended to the Atlantic, Gulf, and Pacific coasts. The productive potential of the city was unparalleled, and the pace of its industrial expansion reached an explosive level.

But the material achievements reveal only part of the picture. It is true that much of the city's life reflected a brutal concentration of naked industrial power, with all the possibilities of corruption, indifference, and cruel exploitation. It is equally true, however, that the intellectual and artistic life of the city flourished with surprising vigor. At the time of incorporation, the life of the community was enriched by newspapers, bookstores, debating societies, and musical programs. By 1865 the citizens had established the Historical Society, the Academy of Science, an art museum and school, and a number of large private libraries, chief of which was the collection of Walter L. Newberry, the founder of the famous library that bears his name. The personal lives of the wealthier and better educated citizens included extensive travel abroad, cultivated taste in the arrangements of domestic life, and a remarkable receptivity to new ideas. Before the end of the century these qualities were to expand into a brilliant flowering of the civic spirit. Chicago became very much aware of itself, conscious of its differences from the East, determined to perpetuate and enhance them, dedicated to creating an original American culture free from stultifying traditions. For a generation or two around 1900 it was marvelously successful.

It is only when we realize the intellectual vigor as well as the industrial power of the city that we can understand how the forms and techniques of contemporary urban building are in good part a Chicago achievement. That accomplishment was bound to come soon or late; the disastrous fire of October, 1871, made it a matter of urgent necessity. The need for a great number of large commercial buildings that were thoroughly functional and that could be erected in the most efficient and economical way was soon to call forth creative ability of the highest order. The first indication of what was to come was the Nixon Building, designed by Otto H. Matz and in process of construction when the fire struck. It represented the most advanced structural techniques available at the time. The walls of the five-story building were divided into slender masonry piers that made it possible to open the elevations into an area of glass nearly equal to that common among the cast-iron fronts. For the interior construction Matz aimed at complete fireproofing. The interior frame consisted of cast-iron columns and girders supporting wrought-iron joists and roof rafters. The upper surfaces of all framing members were covered with a one-inch layer of concrete, and the ceilings were plaster of Paris. Partly by design and partly by luck the structure of the Nixon survived the fire intact, and the building was opened to use a week after the catastrophe.

The Nixon was the first notable example of masonry bearing wall and interior iron frame. As we have seen, it was common enough in the East at the time, but in Chicago it was to lead to daring examples of technical virtuosity even within its own limits. The next step came with the construction of the Leiter Building of 1879, the first important work of Chicago's leading structural innovator, William Le Baron Jenney, who was educated at the École Centrale des Arts et Manufactures in Paris. The Leiter is a curious mixture of the vernacular and the sophisticated: timber floor beams and cast-iron columns form the main structural system; but the distinguishing feature is the location of the peripheral columns immediately inside the piers, which function entirely as fireproof envelopes and as supports for windows and mullions. If the wall columns had been inserted in the piers, and if three more columns had been added to the party wall on the north, the Leiter would have been a work of true skeletal construction. Jenney was to bring together most of the necessary factors for this achievement five years later in the design of his Home Insurance Building.

Meanwhile, Chicago builders were pursuing the possibilities of the masonry-iron combination in a different but equally valuable direction. As we saw in the case of Post's Produce Exchange, a special problem arose from the need to support floors and roofs over large open areas inside the building. The first essay in Chicago was Dankmar Adler's Central Music Hall (1879), a thoroughly traditional masonry exterior that embraced a theater with a roof carried on flat iron trusses; the open volume was surrounded by a shell of iron-framed office floors. A much larger work of the same kind was the Chicago Opera House (1884–85) of Henry Ives Cobb and Charles Frost. The interior theater undoubtedly offered another example of truss framing, although the structural details of the opera house have been lost. These buildings were preparations for the masterpiece of its kind, Adler and Sullivan's Auditorium Building, erected between 1887 and 1889 (Figure 40). The ceiling over the four thousand–seat theater is carried on elliptical trusses with a maximum span of 117 feet; flat trusses support the roof over the theater, the ceilings of the main dining room and banquet hall of the former hotel, and the ceiling of the rehearsal hall above the theater. The remainder of the old hotel and office areas are framed in iron within the granite and limestone piers of the exterior walls. In the Auditorium (now Roosevelt University) Adler combined on an unprecedented scale most of the structural techniques available to him at the time.

Although the masonry bearing wall was already outmoded for high buildings when the Auditorium was begun, it was to be combined once more with the most advanced techniques of iron framing in an architectural classic. The Monadnock Building, designed by Daniel Burnham and John Wellborn Root and erected between 1889 and 1891, rises to a height of sixteen stories on exterior walls of brick 6 feet thick at the base (Fig. 41). By using girders projecting outward from their supports on the masonry piers, the architects were able to open the wall into bow windows with such generous areas of glass as to belie the masonry construction. The interior floor and roof loads are carried on a braced iron frame distinguished by a kind of wind bracing known as portal bracing, a term which indicates the origin of this kind of bracing in the portal frames of truss bridges. In this simple but effective form rigidity is secured by riveting the deep girder to the column throughout the depth of the girder web. The Monadnock and Jenney's Manhattan Building, designed and built simultaneously, appear to share the honor of being the first American buildings constructed with a complete system of portal bracing.

The decisive step in the creation of the framing that made the modern skyscraper possible came five years before the Monadnock's completion. The Home Insurance Building of William Le Baron Jenney, erected between 1884 and 1885, did not embody all the features of the fully developed skyscraper, but enough of the essentials were there to show exactly how this would later be done (Fig. 42). This celebrated building marked the culmination of a century's progress in iron-framed construction. We have already considered the American precedents (pp. 83–85) and commented that Jenney's knowledge of them probably came from George H. Johnson and Peter B. Wight, who knew the structures and whose company furnished the fireproof tile for the Home Insurance. Two buildings in France anticipated the structural system of the Chicago skyscraper. The earlier was the warehouse of the St. Ouen Docks in Paris, erected between 1864 and 1865 after the plans of Hippolyte Fontaine. Though not a skyscraper in the strict sense of the term, the warehouse is unquestionably the first multistory fireproof building in Europe, and possibly the world, in which all loads, including those of the curtain walls, are carried on an iron frame. It was followed in a few years by a curiosity of iron framing, the Menier Chocolate Works at Noisiel (1871–72). The diagonally braced framework of this multistory building, designed by Jules Saulnier, is in effect a large lattice truss that supports the floors and curtain walls.

How much Jenney knew of these various structures has not been exactly determined, but it seems reasonable to suppose that as an architect and engineer trained in a French technical school and as a leader of his profession in Chicago he was abreast of current innovations. The Home Insurance was the first extensive application of the internal skeleton and the curtain wall to a high office building that had been designed to meet the most rigorous functional requirements. The criteria that established Jenney's program were chiefly these: maximum durability and fire-resistance, utmost economy of construction, maximum admission of natural light, and open interior space for maximum freedom in arrangement of internal elements. Skeletal construction alone provided the comprehensive answer to these stringent demands, and by thoroughly exploiting its utilitarian possibilities in both the exterior form and the planning of the large commercial building, Jenney and his fellow architects in Chicago succeeded in creating a new architectural style, the first since the beginning of eclectic revivalism at the end of the previous century. This new organic architecture provided the first modern exemplification of Louis Sullivan's famous dictum that "form follows function."

The most advanced features of the framing system in the Home Insurance were, first, the method of supporting each bay of the wall on a shelf angle fixed to the spandrel girder and, second, the introduction of steel into the building frame. The superior physical properties of the metal compared with those of cast and wrought iron, its greater strength in both tension and compression, and its greater resistance to fatigue had already turned the bridge builder to its use, and by the nineties it was to become a necessity in high-rise buildings. The frame of the Home Insurance included cast-iron cylindrical columns, built-up box columns of wrought iron, spandrel girders and floor beams of wrought iron up to the sixth floor, and Bessemer steel beams and girders above.

Several characteristics of the Home Insurance Building prevented it from being a mature skyscraper, that is, wholly free from dependence on masonry elements. A small portion of the total load was carried on the granite piers at the base of the façade and on the brick party walls at the sides. The connections of the iron and steel frame were bolted rather than riveted, an older practice (now recently revived) which at the time did not provide a joint of maximum rigidity. A puzzling feature was the total absence of wind bracing, an omission that the designers may have defended on the ground that the masonry bearing members and the heavy masonry cladding on exterior columns and girders provided sufficient rigidity for the whole structure. As a matter of fact, the view persisted for nearly a decade that buildings with relatively extensive horizontal dimensions needed no internal bracing. The final element in the Home Insurance that belongs to a day that was passing was the use of raft footings to spread the load of columns and walls widely over a medium hardpan clay about 12 feet below grade. The compressible and water-bearing soil of Chicago offered difficult problems in the construction of foundations that would be proof against excessive settlement. Within a decade after the completion of the Home Insurance, building loads had reached the point where footings had to be placed on piling driven to impenetrable hardpan about 55 feet below grade or on concrete caissons carried to bedrock 110 to 125 feet below grade.

Jenney had marked the path so clearly with his Home Insurance Building that four years were sufficient to bring the steel-framed skyscraper to maturity in Chicago. Each year saw technical innovations that brought it closer to realization. The Tacoma Building of Holabird and Roche (1887–89) introduced the riveted skeleton. The Chamber of Commerce Building of Baumann and Huehl (1888–89) had a steel and iron skeleton that took all the load except for a small proportion that was carried by the masonry envelopes of the columns at the base. The Chamber of Commerce was distinguished by an interior light court extended through the entire height of the building, from the main floor to the gabled skylight in the roof. The corridors then became open galleries running around the periphery of the court and carried on iron brackets cantilevered from the columns. This feature exposed to clear view the multiplicity of the floors together with some of the ironwork that supported them. It was an exciting novelty, and the architects of the time loved to exploit its visual possibilities. E. Townsend Mix, in his design of the Metropolitan Building in Minneapolis (1888–90), sought nearly complete transparency by specifying heavy slabs of glass for the gallery floors (Fig. 43). The Brown Palace Hotel in Denver (1887–92), designed by the Chicago architect Frank E. Edbrooke, offers an even richer example of the same construction. The iron columns and girders supporting the unusually wide galleries around the five-sided court are fully visible, and the light from the gabled skylight at the top of the ninth floor is filtered through a horizontal stained-glass screen carried in a grid of thin iron beams. These exhibitions of structural virtuosity seem extravagant by our standards today, but they were exactly in the spirit of the Chicago school: the courts were dictated by the need for natural light, and their rich ornament was developed organically out of the structural necessities of the iron-framed building.

The ultimate step in the creation of the modern skyscraper came with the construction of the second Rand McNally Building in Chicago between 1889 and 1890. Here the frame of the high office building was wholly freed from its masonry adjuncts and built entirely of steel. The architects were Burnham and Root, and the structural engineers were Wade and Purdy. The presence of these engineering designers reminds us that the day had passed when the architect was the sole master of his craft in all its technical as well as formal details. Once established in Chicago, the fully developed steel frame quickly became standard. Jenney and the engineer Louis E. Ritter designed two big iron- and steel-framed department stores, the Sears Roebuck and the Fair (now Montgomery Ward), both opened in 1891 and both in sound working order today (Figure 44). All the structural features of the modern fireproof high-rise building are here: box columns, girders and beams of I-section, tile arches, portal bracing, riveted joints, fireproof tile cladding, and concrete subflooring. The only elements not present are the little trapezoidal frames that carry the banks of projecting windows or oriels (the iron and steel skeleton of Clinton J. Warren's Unity Building [1891–92] provides a typical example [Fig. 45]).

The steel frame offered unparalleled flexibility in dealing with the structural problems that were associated with buildings designed to house a diversity of functions. The wide range of techniques that may be called upon for a single structure is best exemplified by Adler and Sullivan's Garrick Theater Building, constructed between 1891 and 1892 and demolished seventy years later in a tragic act of architectural vandalism. Originally the German Opera House, the building was designed to include a theater with its associated stage and mechanical equipment, a seventeen-story office tower that extended back over the theater in a thirteen-story rear wing, and club and banquet rooms in the two top stories. The result was a work that presented a great many difficulties of design and construction, from the underpinning of the footings to the framing of the roof. The high load concentrations under the columns of the seventeen-story tower and the proximity of smaller neighbors required unusual precautions in the construction of the foundation work. All column and side-wall footings were sustained by 700 oak piles driven to a dense hardpan located about 55 feet below grade level. Above the piling was a grillage of two layers of oak beams laid at right angles, these in turn supporting a massive raft of concrete reinforced with steel rails and I-beams. The design of this foundation was largely the work of William Sooy Smith, one of the leading bridge engineers of his time, who acted as consulting engineer in the preparation of the Garrick plans. Above the concrete raft in the area of the tower and its two flanking wings were several huge plate girders cantilevered beyond the edge of the raft to carry the columns in the façade and thus prevent their loads from falling directly on individual footings.

The three distinct enclosures incorporated within the building envelope required three separate framing systems differing radically in their details. The office tower and flanking wings were carried by a riveted fireproof steel frame of columns, girders, and joists. Rigidity against wind loads was effected by portal bracing throughout, supplemented by single-diagonal braces in the form of steel rods in certain of the outer bays. The great problem was the support of eight floors of offices above the void of the theater. Adler, as we have seen, had confronted this situation earlier in his design of the Central Music Hall and the Auditorium. In these buildings, however, it was simply a matter of carrying a roof over the theater, whereas in the Garrick it was necessary to support the high concentrated loads of eight stories of columns. Adler again turned to bridge construction for this solution: across the 55-foot width of the theater were six trusses, five of them two stories deep, supported at their ends by the massive brick bearing walls at the sides of the enclosure (Fig. 46). Extremely deep built-up girders carried the proscenium arch and the forward edge of the balcony. Above this assemblage of trusses and girders rose the column-and-beam frame of the upper office floors. At the very top of the rear wing Adler again faced the necessity of covering a large open interior, the club room, without intermediate supports, and once more he adopted truss framing as the solution.

The support of the office floors above the stage involved another set of conditions, which required a third kind of structural system. The rigging loft above the stage rose to the eighth floor level. Since the office areas and the banquet hall above this floor could have no intermediate columns, the floor beams rested on four built-up girders spanning from front to rear of each floor above the stage. But the most remarkable elements in the framing over the stage grew out of a peculiar exigency fixed by the plan: because the office area was slightly narrower than the over-all width of the stage, the brick side walls of this area had to be carried on separate bearing members extending without break through the height of the stage enclosure. These members were two pairs of wrought-iron Phoenix columns, one pair at each side of the stage, rising from the basement to the eighth floor for an over-all height of 93 feet 1 inch.

The foundations and the framing of the Garrick Theater Building constitute an intricate system of construction embracing a great diversity of materials and techniques, but the whole work was the outcome of a careful investigation in the new scientific spirit. It represented the most complex, the most thoroughly studied, and the most advanced structural system available to the builder at the end of the nineteenth century. Preparation for this achievement had been in the making for a long time; yet the specific techniques needed to deal with the problem of the high building had been developed in little more than five years before the Garrick was planned. The structural engineer of our time might find the diversity of elements redundant, but his own highly refined mathematical science was derived in good part from the theories embodied in such works as the Garrick Theater Building.

By 1895 the engineers and architects of Chicago had introduced other innovations to round out the progress of high-building construction in the past century. Adler and Sullivan, again in consultation with William Sooy Smith, provided the final solution to the vexing problem of foundations when they placed concrete caissons extending to hardpan under the west column footings of the Stock Exchange Building (1893–94). The first caissons had been used to support the foundations of the City Hall in Kansas City, Missouri (1888–90), designed by S. E. Chamberlain, but whether Smith knew of this pioneer work has not been determined. The engineer Corydon T. Purdy, in collaboration with William Holabird and Martin Roche, seems to have been the first to use portal arches as wind-bracing when he designed the steel frame of the Old Colony Building in Chicago (1893–94). In this system of bracing the ends of the girder are deepened into quarter-circular fillets, which give the underside of the girder an arched profile. Although the portal provides great rigidity, the depth of the arched girder results in a costly sacrifice of vertical space, so that the net internal volume of the building is considerably reduced for a given over-all enclosure. The Old Colony is another building in which Phoenix columns are the main compressive members. The structural culmination of the Chicago school came with the Reliance Building, erected between 1894 and 1895 after the plans of D. H. Burnham and Company and the engineer Edward C. Shankland. The steel frame of this slender tower of glass is carefully braced in two ways: 24-inch deep spandrel girders provide the usual portal bracing, and two-story columns erected with staggered joints increase the rigidity of the vertical members. The architects of Chicago continued to create their masterpieces of commercial and public architecture, reaching their highest level in Sullivan's Carson Pirie Scott Store (1899, 1903–04, 1906), but the engineers had little more to do than to refine the details of the steel frame until the advent of welding and rigid framing shortly after World War I.

Chapter Eleven | The Iron Railroad Train Shed

A great variety of iron trusses derived from timber prototypes were adapted to the roof framing of mills, factories, warehouses, and theaters in the latter half of the nineteenth century. The special form of the arched truss was at first confined to monumental domes of iron and the domed and vaulted enclosures of exhibition buildings. The most extensive installations of truss framing, however, came with the construction of the train sheds for the big metropolitan terminals that multiplied after the Civil War. The war naturally retarded the construction of railroads, but the end of the conflict brought on a resumption of activity that quickly reached an explosive pitch. By 1900 the railroad system had been increased by 150,000 miles of line, attended by an even greater increase in the mileage of subsidiary tracks in yards, stations, sidings, and extra-track main lines. Economic growth, the multiplication of large cities, increasing personal and social mobility, and the expansion of the nation's land area—these were the primary factors in the rise of passenger traffic to the most profitable branch of rail operations. The major carriers felt compelled to build their metropolitan terminals on a grand scale, huge in size to handle the volume of traffic, extravagant in facilities and decoration to attract the passenger. In the fiercely competitive age that flourished before the passage of regulatory legislation, the principle followed by the railroad companies was to build separate stations in the large urban areas rather than to pursue the obvious economic and civic advantages of combining facilities into union stations. The main result for the history of station building was a variety of structural techniques rather than standardization; in any one city, as a matter of fact, the forms might range from the crudest vernacular construction to the most sophisticated scientific design.

The railroad terminal has always embraced two fundamental parts, the station building, or head house, and the track and platform area. Until the last quarter of the century the tendency was either to incorporate the track layout in the station building or to enclose it in masonry walls continuous with those of the building proper. Such was the case with vaulted sheds on timber trusses, even when the span reached the 166½ feet of Great Central Station in Chicago. As traffic grew, however, the track area expanded so much that its covering became a separate entity. The practice of closing the open end of the shed with an arcaded masonry wall was soon abandoned, although the masonry walls along the sides persisted for architectural purposes until the end of the century. The ruling principle throughout the period was to shelter the tracks and platforms under a single all-embracing roof, preferably without intermediate supports. In the largest cities the growth of suburban service led to an extremely high density of traffic, at times reaching seven hundred trains a day. The great number of short trains and the problem of dissipating smoke established the dimensions of the vaulted shed: the length seldom exceeded 600 feet, the breadth was sometimes as much as half this figure, and the height was one-third to one-half the span.

The timber-and-iron trusses of the 1850's began to give way to all-iron forms in the next decade. The first system of iron trusses supported the shed of the original Cleveland Union Depot, built between 1865 and 1866 by the constituent companies of the Lake Shore and Michigan Southern and the Cleveland and Pittsburgh railroads. The engineering designer was B. F. Morse. The shed rested on shallow wrought-iron Howe trusses curved to the segmental profile of the roof section, their ends carried by masonry side walls. These trusses were tied and braced by an elaborate system of wrought-iron tie rods and cast-iron struts apparently derived from a similar system of ties and braces under the train-shed trusses of the Gare de l'Est in Paris (1847–52). When the shed was demolished in 1914, the engineer in charge of the work found it impossible to calculate the stresses in the various members, with the result that the wreckers had to support the entire structure on falsework as they took it apart piece by piece.

The gable sheds of the next two decades revealed more elegant truss forms. The Baltimore and Potomac Station at Washington, D.C. (1873–77), had a gable shed of wood planking carried on wrought-iron Pratt trusses with arched bottom chords, familiar and determinate forms that could be precisely designed. This frame was one of a long series of arched structures designed by the Philadelphia engineer Joseph M. Wilson for the Pennsylvania Railroad and its affiliates. The largest single-span gable shed was that of the Central of New Jersey Terminal at Jersey City (1887–89). The wrought-iron Pratt trusses carrying the tin-sheathed plank roof spanned 142 feet 7 inches. These were tied together and braced against wind loads by flat longitudinal trusses that also acted as roof purlins.

In the Washington and Jersey City sheds, the trusses transmitted the roof, snow, and wind loads to separate columns or masonry walls at the sides. The action was thus comparable to that of the combination iron-and-masonry structures of buildings. A more efficient form, analogous to full skeletal construction, came from the unification of the two parts into a single framed structure in which the vault was entirely supported on arched trusses of iron or steel whose springing points lay at the track level and whose horizontal thrust was sustained by tie rods rather than masonry buttresses. There were numerous European precedents for this form. Vaulted sheds on iron ribs or trusses appeared in England and on the Continent as early as the 1840's. The most impressive example is the train shed of St. Pancras Station in London (1863–76), the work of George Gilbert Scott and his engineering collaborators, W. H. Barlow and R. M. Ordish. The essential problem was to calculate the optimum depth and spacing of the arched trusses for sustaining roof and snow loads and of the longitudinal trusses for resisting the twisting and buckling action of the wind. With such works as St. Pancras to guide them, the American builders were ready to try their hands by 1870.

The first native balloon shed to be emancipated from masonry adjuncts and from rule-of-thumb experimenting with supporting structures was that of the first Grand Central Terminal in New York. The decision to build this station on the site which its famous successor still occupies grew out of the initial unification of rail lines in New York City. The city had been served since the early 1830's by the New York and Harlem and the New York and New Haven railroads, whose terminals were located on Fourth Avenue at the edge of the lower East Side. Service to Albany was established in 1851 with the completion of the Hudson River Railroad, which extended down the length of Manhattan Island on the far West Side. Two years later a number of small companies between Albany and Buffalo were merged to form the New York Central Railroad. In 1869 Cornelius Vanderbilt began the creation of his rail empire by gaining control of the Harlem company, the Hudson River, and the New York Central and merging the last two into the New York Central and Hudson River Railroad. The decision to build the terminal for the three roads and the New Haven on largely open land at Forty-second Street and Fourth (now Park) Avenue seemed questionable at the time but proved to be remarkably farsighted. The unification of the lines required a great deal of expensive construction: the first step was laying a line on the east bank of the Harlem River to join the Hudson River and the Harlem railroads; there followed in close succession a new swing bridge over the Harlem River, a masonry viaduct over the length of Fourth Avenue above Ninety-seventh Street, and a walled cut from Ninety-seventh to the terminal throat. Multiple-track bridges and rock cuts were costly enough, but they were only preliminary steps to building the extravagant station that was to become the gateway to the city as well as the chief manifestation of the Vanderbilt wealth.

The construction of the station and the approach lines was carried out simultaneously from 1869 to 1871 (Fig. 47). The architect was John B. Snook, and the structural engineer Isaac C. Buckhout. The L-shaped station building embraced the train shed between the two wings of the L, a plan which was to remain the standard practice until the end of the century. The great roof was a cylindrical vault of corrugated sheet iron and glass covering twelve tracks and five platforms with an over-all length of 600 feet and a span of 200 feet. The primary supports were thirty arched Howe trusses of wrought iron, the springing points of their lower ribs connected by wrought-iron tie rods extending under the tracks to take the horizontal thrust of the arches. The upper ribs of the arches sprang from the top of the brick side walls, which were retained for protection and architectural effect, since the compression in the arch, and hence the thrust, was restricted to the bottom rib. The truss between the ribs functioned largely as a stiffening element. In the roof there were three longitudinal light monitors, one of gable form along the crown line, and two on the flanks sloping in the tangent planes. The monitors were carried by thin truss frames superimposed on the main bearing trusses. The rear end of the shed was closed by a fantastic sheet-metal curtain supported by an iron framework and pierced by ten arched openings at the base for the tracks. Surviving prints suggest that there was no wind-bracing within the shed. If this was the case, the metal curtain and frame functioned as a wind screen, a feature that was retained in the later balloon shed in the more advanced form of a lateral wind truss.

In spite of its great size, the first Grand Central Terminal proved to be inadequate for traffic less than thirty years after its construction. From 1899 to 1900 a series of annexes was built along the west side, which expanded the original track layout to thirty-seven tracks. Two years later the New York State legislature passed a law requiring that all trains on Manhattan be operated by electric locomotives after 1910. What seemed to be an intolerable demand at the time led to the decision to demolish the disordered collection of main station and annexes and to replace them with an entirely new installation. The whole complex was progressively demolished during the construction of the present Grand Central Terminal between 1903 and 1913 (see pp. 183–86).

The trusses supporting the Grand Central shed were fixed arches of the traditional kind that had been built successively in masonry, timber, and cast and wrought iron. Although the form is stable and rigid, it has the defect of being an indeterminate structure, with the consequence that abutment reactions can be only approximately calculated. Further, in semicircular and segmental forms the pressure line deviates considerably from the axis of the rib, so that the arch is subject to bending as well as direct compression. The stresses arising from bending forces can be absorbed by a sufficient mass of material, but this practice results in awkward and inefficient forms, especially when the arch is designed as a light framework of iron. These defects can be overcome by introducing hinges into the crown and springing points of the arch. By dividing the arch into free halves, the forces acting on the hinges can be exactly calculated, and each segment can be fabricated so that the lower rib conforms more nearly to the curve of the pressure line.

The three-hinged arch was developed by various French and German engineers between 1857 and 1864 and introduced into the United States by Joseph M. Wilson in 1867, when he proposed its use for the train shed of the Pennsylvania Railroad's Pittsburgh station. The railroad company regarded it at the time as too expensive because of the necessity of building the hinged segments from falsework, and this view prevailed for twenty years. The company finally adopted it for the construction, 1887–92, of its Jersey City terminal, the chief engineer of which was Charles C. Schneider rather than Wilson. The great size of the shed once attracted wide attention, although the terminal's importance declined after the opening of the Pennsylvania's New York station in 1910 (see pp. 178–83).

The Jersey City station was a preparatory essay for the two greatest balloon sheds built in the United States, both constructed simultaneously in Philadelphia and both the work of Joseph Wilson and his associates. The smaller is that of the Reading Terminal at Twelfth and Market streets (1891–93), and the larger that of the third and last Broad Street Station of the Pennsylvania Railroad (1892–93). The train shed of the Reading Terminal is one of only two single-span balloon sheds surviving in the United States (Fig. 48). The other shelters the track layout of the Baltimore and Ohio's Grand Central Station in Chicago (1889–90), which now provides terminal facilities for only ten trains a day. The sheds of the two Philadelphia stations were nearly identical in construction: the plank roof rested on three-hinged wrought-iron arches grouped in pairs; trussed purlins acted as wind-bracing; ties in the form of steel beams passing under the track level took the arch thrust; and an iron framework carried a glass curtain across the open end of the shed (subsequently removed in the case of the Reading station). The Broad Street shed was the largest permanent roof in the world, its span of 300 feet 8 inches exceeded at the time only by the Galerie des Machines at the Paris Exposition of 1889 and the Liberal Arts Building at Chicago's Columbian Exposition of 1893. The Reading Terminal was electrified between 1930 and 1932 and still provides station facilities for a heavy suburban traffic. The history of the Pennsylvania station was violently interrupted on June 10, 1923, when the great shed and the wooden platforms were totally destroyed by fire and the supporting structure was seriously damaged. While the iron arches were demolished and the old roof was replaced by separate platform canopies, the railroad managed to maintain a traffic of 529 trains a day. The active life of the once famous terminal was nearly ended with the completion of the Thirtieth Street Station in 1934, but the building and the track facilities were not finally razed until 1953. The site is now part of the area of the Penn Center development in Philadelphia.

In a few cases the track layout of a station covered so large an area that the single-span shed would have been a prohibitive expense. The greatest of all continuous roofs forms the train shed of St. Louis Union Station, constructed between 1891 and 1894 as the single terminal for all the lines entering the second-largest railroad center in the United States. The architects of the building were Theodore C. Link and Edward D. Cameron, and the engineer of the train facilities was George H. Pegram, the inventor of a type of truss with a polygonal top chord and inclined posts. All the dimensions of the track area reach the superlative degree: the shed is the largest in span and area; the roof shelters the greatest number of tracks and platforms; the track layout serves the largest number of railroads; the connection between the approaches and the terminal tracks is the most extensive railroad wye, and it is also the only one with a double system of interconnecting tracks. The peculiarity of locating the axis of the track layout at right angles to the approach tracks necessitated the double wye and the consequent reverse movement of trains entering the station. Pegram's enormous shed, covering thirty-two tracks and associated platforms, measures 700 feet in length by 601 feet in width. The crown of the roof stands 75 feet above the track level, but the extreme span gives the vault a greatly flattened segmental profile. The area under the shed is divided into five longitudinal bays by parallel rows of columns, to which the load of the tin-sheathed wooden roof is transmitted by inverted Pegram trusses spanning on the transverse line. The wind-bracing has the usual form of trussed purlins extending longitudinally between the trusses. The successive upper chords of any one row of trusses form a continuous curve that matches the profile of the vault section, and the polygonal lower chords, which would ordinarily be in the top position on a bridge truss, are bowed downward in a form suggesting a rope hanging from several supports. This was the effect Pegram intended, but the bewildering multiplicity of framing elements in the vast dark cavern obscures the architectonic power that the shed's great size might have given it. The dimensions of the track layout were determined by a basic traffic of 260 trains a day at the time of construction, plus a heavy mail, express, and transfer traffic.

The only other shed comparable to the St. Louis roof, that of South Station in Boston (1896–99), was replaced by platform canopies in 1934. The true balloon shed on hinged-arch trusses reached the end of its active life with the simultaneous construction of the Pennsylvania and the Pittsburgh and Lake Erie stations in Pittsburgh (1898–1901); the sheds of these stations were also later replaced by platform shelters. Another type of train shed—in the form of low parallel vaults carried on rows of columns that were located on the center lines of platforms—was invented by Lincoln Bush of the Lackawanna Railroad and first used for that company's Hoboken Terminal, constructed between 1904 and 1906 (Fig. 49). The Bush shed enjoyed a considerable popularity for about ten years, when it too fell before the simple and structurally uninteresting canopy. The iron-framed balloon shed and its associated forms, once among the proudest structural achievements of the past century, enjoyed an active life in the United States of only thirty years. The reasons had become obvious almost at the moment when Wilson was carrying the form to its culmination in the two Philadelphia stations. First was the high cost of constructing the arches from falsework, a process which was often carried out by means of traveling platforms designed and operated so as not to interfere with the traffic below. After construction came the expense and difficulties of maintenance. One purpose of the high vault was to diffuse the smoke of locomotives far above the tracks and to dissipate it quickly through ventilating slots, but this was seldom realized because variable winds, accumulated snow, and high traffic density frequently trapped the sulfur-bearing smoke long enough to cause severe corrosion of the ironwork. The need for drastic economies in the depression of the 1930's finished off most of the great sheds; the few that remained did so only because declining passenger traffic and the diesel-electric locomotive eliminated the problem of smoke. For all its short life, however, the balloon shed played an important role in structural history: along with the exposition buildings of the nineteenth century, it was the forerunner of all the open, wide-span enclosures that most dramatically express the spatial character of contemporary building.

Chapter Twelve | Long-Span Bridges in Iron and Steel

All the economic factors that operated to make buildings higher and train sheds wider were also at work to make bridge spans longer. The waterways had always been present, and the railroad lines that had to cross them were rapidly multiplying in the decades following the Civil War. The additional determinants peculiar to bridge construction were the requirements of adequate channel clearance over navigable waterways and the reduction of the cost of pier and foundation work, always the most expensive part of building over wide streams. But the constant lengthening of span would have been impossible without the rapid progress of the metallurgical and structural sciences. By 1865 a variety of valid truss forms were available, especially those invented by Pratt, Warren, and Whipple, and the engineers were well on the way to understanding how these could be adapted to the special conditions imposed by long spans. The physical limitations of cast and wrought iron were soon to be overcome by the widespread adoption of steel after the preliminary essays of the 1870's. Arched and suspended forms in the new metal eventually came to compete with the truss for long-span highway bridges.

The designation of *long-span bridge* as a separate category is somewhat arbitrary, and like the skyscraper, it must be understood in relation to the technical possibilities of the time. A length of 400 feet may be taken as a convenient index for 1865: below that figure the trusses already in use performed well enough; above it the magnitude of forces began to introduce new problems that required special variations on the basic forms. And, of course, the engineer had always to keep in the forefront of his calculations the ever increasing weight of trains, particularly the increasing unit axle load of locomotives. The longer span carried a longer train and hence a greater load; greater loads and deflections meant greater forces of tension and compression in the individual members of the truss. In order to keep the working stresses within safe limits, the truss had to be deepened, the proportions of individual members exactly calculated, and their cross-sectional areas increased in proper degree. Long spans made an old problem crucial: as the general dimensions of the truss expanded, the long web members were subjected to buckling as well as to the direct stresses of tension and compression, with the consequence that additional members had to be introduced into the web system to offset this tendency. Forces other than traffic loads were greatly magnified in long trusses: moderate wind loads reached dangerous levels on the sides of long trains; once negligible factors such as train sway and the horizontal forces associated with braking and accelerating became matters of serious concern. Top and bottom frames, which brace the structure against horizontal forces, had to be greatly strengthened; trusses had to be made rigid in the vertical plane to withstand the complex action of simultaneous swaying, twisting, and buckling forces. Heavy riveted connections had to take the place of the pinned joints of early iron trusses. The volume of piers, footings, and abutments had to be expanded to sustain the increased weight of superstructure and traffic, and at the same time piers had to be carried to greater depth to reach rock or unyielding beds of gravel, hardpan, or sand. The cost of pier and foundation work reached the level where the time-tested masonry had to be abandoned in favor of the new material, poured concrete.

For a variety of historical, geographic, and economic reasons, it was again the Ohio River that provided the challenge and the opportunity to the builder of long-span bridges. The earlier railroads that eventually formed the trunk lines of the eastern and southern systems had to cross the river in order to unite the networks of the eastern, southern, and Great Lakes regions. For these spans the Whipple truss was dominant until about 1890, although the innovations that were eventually to supersede it were being introduced at the same time. The first step in the direction of long-span construction was the Ohio River bridge of the Pittsburgh and Steubenville Railroad, erected at Steubenville, Ohio, between 1863 and 1865 (Fig. 50). Jacob H. Linville of the Pennsylvania's engineering staff was the chief engineer. He used a variation of the Whipple truss for the river crossing, with modifications that he introduced himself for greater strength and rigidity. The 320-foot length of the channel span set a new record for the time, but this was soon to fall. The Steubenville bridge was replaced between 1904 and 1906, but a model of the channel span is now a permanent exhibit in the Smithsonian Institution.

The span length of the Whipple truss grew rapidly in the next decade, first exceeding 500 feet in the Ohio River bridge of the Cincinnati Southern Railway, constructed at Cincinnati from 1876 to 1877. The engineers in charge of this important project were Jacob Linville and Louis F. G. Bouscaren, chief engineer of the railroad company. Cincinnati was the natural gateway between the Great Lakes region to the north and the coal fields and agricultural areas of the South. Until 1872, when the first railroad bridge was completed, only Roebling's suspension bridge united the city and the Kentucky shore, and it was restricted to highway traffic. The municipal government recognized the necessity for a second rail span and took the steps to realize it. Seven years later the city completed construction of a railroad line from Cincinnati to Chattanooga, Tennessee, and in 1880 leased it for operation to the Cincinnati, New Orleans and Texas Pacific Railway. The latter company is now controlled by the Southern, but the city still earns a rental from its rail property.

The Ohio River bridge was the major structural undertaking in the building of the railroad, which is known for the number of bridges and tunnels along its route through the Cumberland Mountains. The 515-foot length of the channel span established another record that was to last for more than a decade. This drastic change from the standards of pre-Civil War practice was dictated mainly by the need to clear the concentrated steamboat traffic at the Cincinnati waterfront. The most important feature of the whole project, however, was the specifications that Bouscaren prepared for the contractor. This document is the prototype of all subsequent bridge specifications, for it was the first to embody in detail all the criteria of design, loading, material, workmanship, and safety that the builder follows in the erection of a major railroad span. Indeed, the whole modern program of construction was prefigured in the Cincinnati project—geological and hydrographic survey of the site, determination of loads, stress analysis and calculation of the size of truss members, preparation of working drawings and specifications, competitive bidding, the testing of sample materials, inspection at the mill and at the site, and the training of mill hands and construction workers. With this program as a guide, the builder could work with confidence and the public could expect an end to the grim series of disasters that once marked the history of the iron bridge.

The exclusive use of cast and wrought iron for the truss bridge became a thing of the past two years after the completion of the Cincinnati bridge. William Sooy Smith first specified steel for all structural members in his design of the Chicago and Alton's Missouri River bridge at Glasgow, Missouri, constructed between 1878 and 1879. The five spans of the river crossing were composed of Whipple trusses with a maximum length of 314 feet. In spite of the great gain in strength with little increase in weight, the Glasgow bridge was replaced ten years after it was completed because of the sheer weight of traffic. The new metal was used on a much larger scale in the Illinois Central's Ohio River bridge at Cairo, Illinois (1887–89). The designing engineers, George S. Morison and Alfred Noble, specified wrought iron for the diagonal members of the Whipple trusses and steel for all others. The Cairo bridge brought the Whipple truss to the end of its active life. The chief reason for its abandonment was the extreme length of the double-panel diagonals for the 519-foot channel trusses. Any increase in the over-all dimensions of the truss would result in excessive flexibility in the diagonals with a consequent loss of rigidity in the individual panels.

Not only did the single-panel diagonals of the Pratt and Warren trusses offer greater rigidity for long-span bridges, but this inherent advantage could be improved by subdividing the panel in a way that would minimize buckling and distribute the loads more uniformly while retaining the determinacy of the original form. Subdivision by means of half-length hangers added at the midpoints of the panels was introduced by Albert Fink in a truss design patented in 1857. Subdivision by adding half-length diagonals was introduced by Montgomery C. Meigs in 1858 for the spandrel trusses of his Pennsylvania Avenue bridge in Washington, D.C. (for the arches of this bridge, see p. 94). The idea was first applied to the truss bridge when Albert Fink designed the structure erected by the Pittsburgh, Cincinnati and St. Louis Railroad over the Ohio River at Louisville, Kentucky, between 1868 and 1870. The 400-foot channel span was different in form from the other twenty-six spans in this mile-long bridge. For the channel trusses Fink developed a modified form of the Warren truss in which the individual panel was subdivided by an intermediate post and a half-length diagonal added to the lower half of the panel. The arrangement of cast- and wrought-iron members followed the usual distribution of compressive and tensile stresses. Fink introduced the subdivisions in order to distribute the load of the compressed top chord as uniformly as possible through the posts and diagonals, to minimize buckling in the full-length diagonals, and as a corollary, to reduce the stresses in the joints.

Methods of subdividing the Pratt truss similar to Fink's modification of the Warren were developed by engineers of the Pennsylvania Railroad in the decade following the opening of the Louisville bridge. The Baltimore truss, with a flat top chord, came in 1871, and the Pennsylvania truss, with a polygonal top chord, in 1875. A somewhat more complex form of the Pennsylvania truss in which an intermediate horizontal strut was added to each panel was designed by William H. Burr for the Ohio River bridge of the Chesapeake and Ohio Railway at Cincinnati (Fig. 51). Built between 1886 and 1888, the bridge was the second rail span at Cincinnati and the one that brought the company's main line into the city. The channel span established a new record of 545 feet, which was destined to fall in seven years. A new railroad bridge was built between 1928 and 1929, but the old structure is still in service for highway traffic.

The subdivided Pratt truss was extended another two feet in length when the Cleveland, Cincinnati, Chicago and St. Louis Railroad (Big Four Route) built its Louisville bridge between 1889 and 1895 (replaced in 1929). In retrospect, we can see that these huge spans mark a mature level in the science of bridge design since this modification of the truss form has been employed with little change up to the present time. That is more than we can say, however, for the process of construction. The erection of the Big Four bridge cost thirty-seven lives as a consequence of the combined action of a storm and two exhibitions of gross carelessness. One truss fell into the water because the builder had provided inadequate bracing and underpinning for the falsework, and another was overturned because the construction crew failed to make a secure connection between the new and the finished span. The long-overdue result of these disasters was that the problem of safe work practices became the object of scientific attack, with the purpose of reducing the hazards associated with the construction of a big bridge to a minimum. They have seldom been wholly overcome.

Before the end of the century the very longest truss spans had come to be built in combinations of two cantilevers and a suspended span. The bridges at Cincinnati and Louisville that we have just described were composed of two parallel rows of simple trusses, that is, trusses supported at their ends. The cantilever beam or truss, on the other hand, is supported at only one end and hence must be anchored in place by a second truss that extends in the opposite direction from the pier. The action of a cantilever is in some ways the opposite of that of a simple member. When a simple truss is deflected under load, it bends downward in such a way that the lower surface is convex (positive bending); the cantilever, however, bends in the opposite way, so that its upper surface is convex (negative bending). This means that the distribution of tension and compression in the chords of the truss is reversed: in a simple truss the bottom chord is in tension and the top chord in compression, whereas in the cantilever the stresses act in the opposite way, the top chord being in tension and the bottom in compression. The distribution of the bending moment is also reversed: in the simple truss it is a maximum at the center, but in the cantilever it is a maximum at the anchored end, which is over the pier or abutment. The arrangement of web members is essentially the same in the two types. No cantilever bridge of truss construction appears to have been built in the United States until after the Civil War (for the earlier history, see p. 57). In 1867 the Solid Lever Bridge Company of Boston proposed the construction of iron-truss cantilevers with web members arranged as in the Warren truss. The designer was C. H. Parker, who claimed to have built several such bridges in New England and New Brunswick, but the record is unreliable.

The first railroad cantilever bridge in the United States was built by the Cincinnati Southern Railway to carry its line over the Kentucky River at Dixville, Kentucky. It was constructed from 1876 to

1877 on the basis of a design jointly developed by Charles Shaler Smith and L. F. G. Bouscaren. Since flash floods of disastrous proportions made it dangerous to erect the trusses from timber falsework, Smith planned the bridge as a parallel pair of continuous Whipple trusses supported on the two abutments and on two intermediate iron bents rising from masonry footings at the shores of the stream. A continuous member is indeterminate and its pattern of deflection is complex: in the region of the supports it is subject to negative bending, like the cantilever; between the supports, however, it is deflected positively, like the simple beam. At the points where the deflection changes from negative to positive (points of contraflexure), the bending is zero. The theory of continuous members had been previously developed to a fairly mature state by European scientists, among them Karl Culmann, whose American visit we discussed in connection with the timber truss (pp. 59–60).

But the continuous truss was still a novelty in the United States in the seventies and its construction offered certain difficulties, chief among them the high secondary stresses arising from the settlement of the piers. Its primary virtue is that a load at any one point can be distributed throughout the entire structure—so that all the members share the burden, so to speak. In order to enjoy the advantage and minimize the risk at the Dixville project, Bouscaren proposed that hinges be inserted in the chords of the trusses at the points of contraflexure between the abutments and the intermediate supports. The precedent for this idea was a bridge built in 1867 over the Main River at Hassfurt, Germany, which was designed by Heinrich Gerber on the basis of Culmann's theory. By locating the hinges at the points of zero bending, the action of the structure closely approximated the theoretical calculation of stress, and at the same time the high stresses resulting from pier settlement were largely prevented. The presence of the hinges transformed the continuous truss into a complex structure consisting of cantilevers and fixed spans. The center span, over the water, could then be erected by cantilevering the two halves outward from the fixed ends of the permanent cantilevers until the halves met at midstream. It was a bold idea, appropriate to the rugged and isolated region of the Kentucky gorge—"a work of science without concession," as Louis Sullivan said of it. Yet certain features of the bridge make us realize that it was a pioneer work of immature form: the uniform depth of the truss was uneconomical in that it did not reflect the changing bending moment, and the location of the cantilevers between supports and abutments was less efficient than to face them in the opposite direction, which would have allowed a longer clear span between supports without any increase in the maximum depth of the truss. The Dixville span performed satisfactorily, however, until heavier traffic forced its replacement in 1911.

Maturity came rapidly in the case of the cantilever, for it was introduced at a time when the science of bridge design had reached a high stage of development. The first large bridge to embody the now universal principle of using the cantilevers to support a suspended span between them was the Niagara River crossing of the Michigan Central Railroad at Niagara Falls. This double-track structure, with a 470-foot central span, was built in eight months during 1883 under the direction of Charles C. Schneider. It was replaced in 1925 by the massive steel arch that now serves the railroad. The construction of the rail bridge over the Firth of Forth in Scotland between 1883 and 1890 provided a powerful stimulus to the development of the cantilever. The incredible dimensions of this bridge—a clear span of 1,700 feet and a truss depth of 330 feet—suggested unlimited possibilities to the engineer. The immediate consequences were a rapid increase in the size of trusses and the substitution of steel for iron in the primary structural members. A new American record was established in 1888 with the completion of the Hudson River bridge of the Hartford and Connecticut Western Railway at Poughkeepsie, New York. The channel span extends 548 feet through a combination of two cantilevers and a suspended span. The Poughkeepsie bridge once enjoyed a number of distinctions—its one-mile length made it the longest continuous steel bridge, its concrete footings were the largest cast up to that time, and as the only bridge below Albany it provided the first direct link between New England and the eastern coal fields. The bridge still serves in this capacity, unchanged except for the reduction of the original two tracks to one to avoid eccentric loading.

The year the Poughkeepsie bridge was completed saw the beginning of work on the cantilever span over the Mississippi River at Memphis, Tennessee. Built by the Kansas City, Fort Scott and Memphis Railroad, it was the creation of two leading talents of the day, George S. Morison and Alfred Noble (Fig. 52). The aim was again to connect two fast-growing rail networks separated by a broad waterway. The need was particularly acute at Memphis, since there was no bridge over the Mississippi for the entire length of the river below St. Louis. The natural obstacles provided enough serious difficulties, but these were compounded by the bitter opposition of the steamboat interests. When the Rock Island Railroad built the first bridge over the river in 1856, the steamboat companies made their hostility clear by arranging a collision that set fire to one of the spans. Thirty-five years later they were still implacable but were at least resorting to legal means. They demanded a channel span 1,000 feet long, which would have made the necessary truss construction prohibitive in cost. A special board of engineers appointed by the federal government to inquire into the matter concluded in 1888 that a span of about 800 feet ought to satisfy all the parties concerned, and the engineers of the bridge based their design on this figure. The actual length of the channel crossing in the finished steel structure is 791 feet, a span that was not to be exceeded for American railroad bridges until the completion of the Hell Gate arch in 1916. The Memphis bridge is an irregular combination of cantilevers, anchors, and suspended spans, a pattern dictated by the position of the channel and the shore topography. The west approach is an impressive structural feat in itself: the low-lying, easily flooded land on the Arkansas shore required that the tracks be elevated on a mile-long steel viaduct of Warren deck trusses.

The Memphis project carried the railroad truss to its greatest clear length, with the result that subsequent cantilever structures seem anticlimactic by comparison. The one exception was a bridge whose spectacular character arose from its setting as much as its size. The extension of the Galveston, Harrisburg and San Antonio Railroad westward to El Paso, Texas, brought the track to the deep limestone gorge of the Pecos River in 1891. In little more than a year the builders put up the high span that carried the single-track line like a stretched thread over the water far below it (Fig. 53). The bridge stood 328 feet above water level, but its central cantilever-suspended span extended only 217 feet 6 inches between supports. As originally built, the structure consisted of lattice deck trusses flanking the irregular Warren trusses of the longer river and shore spans; plate girders replaced the lattice members in 1910. The high slender bents and the thin trusses, standing against the bare rock masses of the West Texas desert, gave the bridge a delicacy that may have had more aesthetic appeal than the huge trusses of long-span bridges. The Pecos span was entirely replaced in 1944 by what is unquestionably the most elegant railroad truss bridge in the United States (see pp. 217–18).

Truss construction had come of age by the end of the nineteenth century after a progressive evolution that extended over fifty years; the steel arch was to do so in one bold leap at St. Louis. Few iron arches were erected in the formative period before the Civil War. Meigs's use of iron water mains as arch ribs in his Washington bridge (1858) was an ingenious idea, for the hollow tube provided a structural form of maximum efficiency, and the casting of relatively short segments of iron pipe was a well developed technique. A promising advance in the evolution of tubular-rib arches came with the construction of the Chestnut Street Bridge over the Schuylkill River at Philadelphia between 1861 and 1866. Six parallel ribs carried the deck over a clear span of 185 feet. As far as the record shows this was the third iron arch built in the United States. (The original one was constructed at Brownsville, Pennsylvania, in 1836; see p. 94).

On the basis of this narrow precedent, James B. Eads proposed in 1867 to span the Mississippi River at St. Louis with a double-deck arched structure generous enough in clearance to offer minimum obstruction to river traffic (Fig. 54). There had been four proposals to unite the city with the Illinois shore since 1839, among them two suspension bridges respectively suggested by Ellet and Roebling. Eads, who had no training as an engineer and no formal education after the age of thirteen, finally persuaded the business community of St. Louis to accept his plan. What he lacked in education was more than balanced by other virtues and abilities, chief among them courage, a natural inventive faculty, high standards of integrity and workmanship, and a sound judgment of the characters of others. Born in Lawrenceburg, Indiana, in 1820, he began his long and intimate association with the Mississippi River in 1838 as a purser on a steamboat. In 1842 he invented a diving bell from which he made a fortune salvaging sunken ships. He established a glass manufacturing business in St. Louis that brought him further prosperity and a good civic reputation. He invented and built an armored steam-propelled gunboat, which proved valuable to the Union Army in the Mississippi campaign. His first-hand knowledge of the river was the primary factor in the acceptance of his daring scheme and in his appointment as the chief engineer of the project. Aware of the limitations of his knowledge, he wisely sought the advice of first-rate engineers and mathematicians. Eads began his work by organizing a staff that included Charles Shaler Smith as chief engineering consultant, Henry Flad and Charles Pfeifer as assistant engineers, and William Chauvenet, chancellor of Washington University, as mathematical consultant. He originally sought the collaboration of Jacob H. Linville, but this talented engineer was the first to refuse association with the project. The refusal was a blow, but the choice of Smith, who had been trained under George McLeod and Albert Fink of the Louisville and Nashville, was a brilliant stroke.

Clearance of the site for the construction equipment and the building of the caissons began in 1867. Early in the following year Eads decided to use steel for the tubular ribs and thus became the first to introduce the metal into American construction practice. His initial choice was plain-carbon cast steel, but after a sample of the metal failed in the testing machine, he turned to the costly chromium steel, a tough, corrosion-resistant alloy that requires no protective covering. The most difficult part of the whole work was founding the piers on bedrock, 86 feet below mean water level for the St. Louis pier and 123 feet below for the Illinois. The novel and little tested pneumatic caisson offered the only feasible solution. The idea of an air- and water-tight enclosure for the construction of subaqueous works was given its first practical demonstration in 1830, when the Englishman Thomas Cochrane used the device for laying up masonry footings in marshy ground. A cast-iron cylinder with a sharp cutting edge at the bottom and sealed sides and top was patented in England in 1843; six years later John Wright used a variation on it for the construction of the Medway Bridge at Rochester. Eads's caisson was an iron-shod timber cylinder 75 feet in diameter, with an 8-foot deep working chamber in the lower end. The masonry of the pier was built up within the cylinder, the increasing weight of the successive courses forcing the cutting edge further into the stream bed. It was a dangerous operation at the time, and twelve lives were lost as a result of the "caisson disease" that follows from drastic changes in air pressure. There were other hazards—tornadoes, ice, and floods—but the work was pushed forward until the piers and the masonry-arch approaches were completed in 1873.

The steel tubes that constitute the arch ribs were erected by the method known as cantilevering under backstays: the completed segment of the arch was held in place against the pier by cables that extended from the free end over towers on the pier or shore abutment to an anchor imbedded in the masonry of the approach. The stays held the lengthening segments of the two half-arches as cantilevers until the arch was closed at the crown. The arches were built simultaneously in both directions from the pier faces so that the opposing horizontal thrusts canceled each other. The arch structure of Eads Bridge is unique. It consists primarily of four vertical pairs of tubes, the pairs set side by side across the width of the deck. All arches are fixed at their ends, the spans of the three being 520 feet for the center and 502 feet for the flanking arches. Each tube, 18 inches in diameter, is built up in 12-foot lengths of six cylindrical segments like barrel staves, which are bolted together through longitudinal flanges. There are six thousand staves in the twenty-four tubes that comprise the three systems of arches in the river crossing. The tubes are joined and braced by a dense system of diagonal bracing in the longitudinal and transverse planes. The two decks of the bridge are carried to the arches by wrought-iron spandrel posts braced only in the transverse planes. The upper deck is a four-lane street and the lower the double-track line of the Terminal Railroad Association of St. Louis. Eads's engineering staff designed the structure for a total load of 28,972 tons, although the maximum possible load that can be placed on the bridge is about 7,000 tons. Now in its tenth decade of service, this enduring structure has had only routine repairs and maintenance since its completion; yet it still carries, in addition to vehicular traffic, the East St. Louis freight of the Terminal Railroad and the passenger trains of the B. and O., Illinois Central, and L. and N. railroads.

Eads adopted the fixed arch for the Mississippi bridge because of its familiarity and rigidity, but there was need for a determinate structure that could be adapted to a wide variety of situations where asymmetrical and other unusual forms might be useful. The answer, as we have seen, was the three-hinged arch, which saw its first American use in the bridge designed by Joseph M. Wilson to carry the Pennsylvania Railroad tracks over Thirtieth Street in Philadelphia. Built between 1869 and 1870, the bridge had a multitrack deck supported by twelve parallel wrought-iron ribs spanning 64 feet clear. The load was carried to the ribs by spandrel trusses in the form of the Pratt truss with pinned connections. Wilson chose the three-hinged form, not only because it is a determinate structure, but also because it offered greater flexibility to compensate for secondary stresses caused by thermal expansion and contraction.

But the three-hinged arch, like the suspension bridge, proved to be too flexible for rail traffic and was subsequently confined to highway bridges. For railroad structures and long-span vehicular bridges the two-hinged arch, although an indeterminate form, offered some of the advantages of the three-hinged type. The two-hinged arch leaped into prominence with the construction of the great steel arch of Gustave Eiffel's Garabit Viaduct in France (1880–84). The size of this graceful arched truss was soon matched in the United States by Washington Bridge, crossing the Harlem River at 181st Street in New York (Fig. 55). Built between 1887 and 1889, this handsome structure, an American classic of steel arch construction, was designed by the engineer William R. Hutton and the architect Edward H. Kendall. The deck of each span rests on six parallel two-hinged arched girders with a clear span between hinges of 508 feet 9 inches. The bridge now carries the traffic of an important crosstown route connecting George Washington Bridge with the Bronx parkways.

The two-hinged arched truss of relatively light construction suitable for vehicular traffic offered the possibility of increasing the length of a single span well beyond that of the conventional railroad truss. The ultimate American record for the past century came with the construction of the highway bridge over the Niagara River immediately below the Falls. Erected between 1895 and 1898 after the design of L. L. and R. S. Buck, the structure was supported by a pair of hinged arches with a span of 840 feet. The web system was that of the Pratt truss. The profile of the arched truss clearly indicated its two-hinged form, since the two ribs were drawn to a point at the hinges. The disaster that struck this bridge was common in the days of timber construction but was thought to be highly improbable in the case of steel. In January, 1938, an immense ice jam in the semicircular embayment carved out by the Canadian falls was funneled through the narrow gorge at the bridge site. The contraction squeezed the ice to a greater depth, and on January 27 it reached the level of the hinges. The irresistible mass sheared the hinged connections, which caused the whole framework of steel to collapse onto the surface of the ice. Buck's span was replaced by the present Rainbow Bridge, which is another structure of two-hinged arched girders.

The wire-cable suspension bridge with masonry towers, braced against wind as well as traffic vibrations, had been brought to maturity by John Roebling when he completed the Cincinnati span. Another of the Roebling creations—the longest and most famous—Brooklyn Bridge—takes us into the era of steel construction and the long-span bridge (Fig. 56). Proposals to span the East River between Manhattan and Brooklyn (a separate city until 1898) had been published from time to time since 1811, when Thomas Pope offered the first plan. No one attempted to translate these ideas into action until the winter of 1866–67, when the weather reached such severity that ferry service was disrupted and Brooklyn virtually isolated from New York. By this time John and Washington Roebling had established reputations that promised results, even in the face of the dangers of trying to span the East River. The New York Bridge Company was organized in 1867, and the directors appointed John Roebling chief engineer in the same year. The unchallenged master of his craft, now in his sixty-first year, moved with characteristic energy and courage. Plans were prepared, the shore sites cleared in 1869, and the caisson for the Brooklyn tower completed in January, 1870. But before construction could start, tragedy ended the greatest bridge-building career in the nineteenth century. A tugboat rammed the piling on which John Roebling was standing and crushed his foot, part of which had to be amputated. He died on July 22, 1869, from the resulting tetanus infection. Fourteen years later his finished monument was there, for all who required one. The directors of the bridge company paid him the highest tribute when they appointed Washington Roebling to succeed him as chief engineer.

The difficulties presented by the East River were much like those of the Mississippi at St. Louis, although the geological character of the two is entirely different. The East is a tidal stream with swift and turbulent currents which carry an extremely dense and diversified traffic. On the other hand, bearing rock lies nearer the surface, markedly so on the Brooklyn side; the river is not subject to the Mississippi's floods, nor does it have the larger stream's shifting and treacherous bed. The construction of the tower foundations for Brooklyn Bridge was a somewhat less formidable task than the building of the piers for the Eads span. The Brooklyn caisson had to be sunk to a depth of 44 feet 6 inches below mean water level, and the Manhattan caisson to 78 feet. In spite of the numerous hazards, masonry work progressed steadily without a fatal accident until the towers and anchor piers were completed in the summer of 1876. Among those who were injured, the chief casualty was Washington Roebling, who was partially crippled by pressure changes for the remainder of his ninety-year life.

With the towers and the anchors in place, the auxiliary elements necessary for the cable spinning could be installed: auxiliary cables, plank footbridges, storm cradles, and carrier ropes followed in that order. The first traveler rope was carried across the river in a scow, and the second spliced into the first to form a continuous loop. Cable spinning was accomplished by means of two sheaves suspended from the traveler ropes, which moved back and forth between the anchors and over the saddles near the top of the towers. For each of the four cables the moving sheaves formed nineteen strands of 286 steel wires per strand. The high-tensile steel marked the second use of the new metal in an American structure. Once the main cables were secured to anchors and saddles, the remaining parts of the bridge could be readily installed in order from the cable downward: first the wire-rope suspenders and radiating stays, then the stiffening trusses and the deck frame, and finally the deck itself, with its elevated walkways and streetcar lines. The deck structure was completed in 1881, and the whole span opened to traffic in the spring of 1883. There was only one fatal accident in the work above the water surface: on June 14, 1878, a strand of cable fell into the river, killing two men and injuring three—a high enough cost but a safety record that has seldom been improved for construction projects of comparable magnitude. The finished bridge outdistanced all others for length of clear span and length of continuous structure. The main span extends 1,595 feet 6 inches between towers, and the total length of deck between anchors is 3,455 feet 6 inches.

For the remainder of the century lesser men than John Roebling built lesser bridges. Washington Roebling turned to the cable-manufacturing enterprise that his father had founded at Trenton, and other engineers introduced the innovations that rounded out the progress of suspension-bridge design. Chief of these are towers of riveted steel frames rather than the expensive masonry; simplification of the suspension system, deck frame, and stiffening trusses; and the reduction of the cables to a single pair. Concrete took the place of masonry for the piers and the anchors, and deepened stiffening trusses provided sufficient aerodynamic stability that Roebling's complex system of stays could be abandoned. All these elements are embodied in the second East River span, the Williamsburg Bridge, constructed between 1898 and 1903 as the chief work of the engineer Leffert L. Buck. The next great age of the suspension bridge was to come around 1930, when the Hudson at New York and the waterways around San Francisco were at last spanned.

Chapter Thirteen | Plain Concrete Construction: Buildings, Bridges, and Dams

Concrete is the oldest synthetic material used in the building process, and yet there was a period of about thirteen centuries when the knowledge of it was wholly lost in Europe. It was used by Mayan builders as early as the eleventh century, but this work had no influence on later building in the New World. Of the four primary materials in concrete, the sand and gravel (aggregate) are inert, and the cement and water are the active ingredients in the complex chemical reactions that transform the wet plastic into a rigid and durable substance. Cement is essentially quicklime (calcium oxide), which can be obtained most easily by burning finely broken limestone. This process, known as calcining, originated in remote antiquity, the earliest known lime-burning kilns dating from about 2450 B. C. Sometime in the late republican period, the Romans made two extremely important discoveries. They first found that ordinary cement will set under water if the lime is previously mixed with a small quantity of a volcanic earth now called pozzolana, after the town of Pozzuoli, Italy. Somewhat later they discovered that if hydraulic lime, sand, and water are mixed with an aggregate of broken stones or bricks and allowed to set, the resulting product is a strong and durable stonelike substance that will harden in water as well as in air. The Roman builders gave the invention its first practical demonstration in the walls of the colony of Alba Fucens, constructed during the third century B.C. During the early imperial period concrete was used for walls, domes, vaults, and breakwaters, and eventually for all structural elements except columns.

These valuable arts were lost in the Middle Ages. Hydraulic-lime mortar was rediscovered in the sixteenth century, when English builders began to import pozzolana and lime mixtures from Italy. But concrete in its Roman form was not revived until the latter part of the eighteenth century, when the English again led the way. Around the mid-century George Semple mixed hydraulic lime and gravel to make a primitive concrete for the foundations of the Essex Bridge over the River Liffey in Dublin. The famous British engineer John Smeaton used a similar concrete in 1760 for the cores of lock walls on the River Calder. Before concrete could be employed extensively, however, a method of providing a steady supply of reliable hydraulic cement had to be developed, and this could be accomplished only through the discovery of a ready supply of the hydraulic agents. These various steps form a complex pattern of evolution in the late eighteenth century that by 1800 had produced experimental knowledge of what was essential in the ingredients.

The discovery of various limestones that produced a naturally hydraulic lime showed that the presence of clayey materials and shales, as well as pozzolana, imparted the hydraulic property. In 1796 James Parker of Northfleet, England, invented a commercially useful method of manufacturing natural hydraulic lime by calcining a shale-bearing limestone found on the Isle of Sheppey. But methods such as Parker's all depended on a natural supply of the necessary ingredients. By 1811 James Frost in England and Louis J. Vicat in France had surmounted this limitation by calcining a carefully proportioned mixture of chalk (relatively pure calcium carbonate) and clay. Although this process marks the beginning of artificially prepared cement, the decisive step for the building arts was not taken until 1824. In that year Joseph Aspdin of Leeds, England, obtained a patent for an artificial cement again prepared by calcining a mixture of chalk and clay, but he carried on the process at a temperature high enough to vitrify the mixture, which was then ground to a powder. He was thus able to produce a cement superior to and more reliable in quality than those of Frost and Vicat. Aspdin called it Portland cement because of the similarity of its appearance to a limestone found in the region of Portland, England. It has retained the name to the present time and continues to be the prime cementing agent in modern structural concrete.

The systematic application of concrete to structural needs did not come in the United States until the latter half of the nineteenth century, but there were sporadic preparations that extended back to the colonial period. The chief event in this irregular history occurred in 1818, when the canal engineer Canvass White discovered a source of natural hydraulic cement near Sullivan, New York. The "water lime" which White prepared and patented in 1819 was derived from local deposits of a clayey magnesian limestone called cement rock. Although he used it originally for facework on walls and aqueducts of the Erie Canal, by the mid-century it had been adopted throughout the country for many kinds of structures. In the thirty years following White's patent at least a dozen discoveries of natural cement were made in the seaboard states and in Kentucky and Illinois. The best of these natural products was the variety found near Rosendale, New York, in 1828 by the builders of the Delaware and Hudson Canal. Rosendale cement, as it came to be called, proved to be superior to White's product and consequently came to dominate American production until the establishment of an artificial cement industry in 1871.

Although enterprising builders began to see the enormous advantages of concrete in the 1830's, construction in the new material was to be greatly hampered by dependence on natural sources of hydraulic lime. Limestone is an extremely abundant rock, but the kinds suitable for the preparation of hydraulic lime are rare. Moreover, natural cement is an impure variety and hence unpredictable in quality and behavior. Construction of concrete bearing members appears to have been initiated around 1830 by Obadiah Parker of New York City. Exactly what kind of cement he used is not clear, but it is possible that he prepared his own product by mixing limestone and clay in the manner of Frost and Vicat. He built a Greek Revival house of concrete in 1835 in which the side walls, entablatures, and cornice were cast as a unit and the columns as separate members. But monolithic construction of this kind was to remain rare until late in the century. The more common technique was to precast the concrete in blocks, which were then laid up in mortar like masonry. The originator was probably G. A. Ward, who built his own house in this way on Staten Island, New York, in 1837. Both these pioneer works reveal the early dependence of concrete construction on traditional masonry forms, a dependence which was to persist until the end of the century.

For the next thirty years structures of concrete continued to be isolated essays that showed little progressive development. Joseph Goodrich built a concrete house at Milton, Wisconsin, in 1844, using Portland cement imported from England. It provided a good test of the durability of the material, for it lasted one hundred four years before it collapsed in 1948. Concrete pier footings for bridges came with the construction of the New York and Erie's Starrucca Viaduct in 1848. Piers of cast-iron drums filled with concrete appeared about the same time, reaching their maximum size in the Missouri River bridge of the Union Pacific Railroad at Omaha (1868–72). The manufacture of precast block on a commercial scale was inaugurated by George A. Frear at Chicago in 1868. The business flourished, for the block was much cheaper than masonry, though inferior to it in durability and strength. Artificial stone, as it was then called, was sometimes cast with pigments in imitation of various natural stones. It was very popular for a time chiefly because it made possible the most elaborate ornamentation at a low cost.

The great age of concrete began in 1871, when David O. Saylor patented an American equivalent of Portland cement and built a mill at Coplay, Pennsylvania, to manufacture the product. This marked the establishment of the artificial cement industry in the United States as well as the beginning of a scientific understanding of the physical properties and structural behavior of concrete. Scientists were soon to learn that the active ingredients in the clays necessary for the production of hydraulic lime are the oxides of aluminum and silicon, although the full understanding of the chemical reactions that occur in the setting of concrete was not to come until well into the twentieth century. The success of Saylor's business was assured when James B. Eads specified his cement for the South Pass jetties at the mouth of the Mississippi River (1875–79). They were built to extend the main shipping channel into the Gulf in order to prevent the deposit of sediment at the point where the channel current loses its velocity as it flows into the sea. The jetties are a remarkable example of megalithic construction, in which some of the blocks poured at the site weigh as much as 260 tons. Concrete is the only material of sufficient density that can be formed into blocks large enough to withstand Gulf Coast hurricanes.

By 1880 concrete, either poured in mass or cast as blocks, had come to be used for a wide variety of simple structural elements such as piers, walls, and footings that are subject only to compression. This restriction arose from the fact that concrete, like stone, has a high compressive but a negligible tensile strength. Builders had learned that the most economical form of the material is the homogeneous mass poured into wooden forms. Since concrete is plastic before setting and since formwork can be built in a variety of shapes limited only by the carpenter's skill, there was no reason why it could not serve for the compressive elements of a large multistory building. The pioneer work in this respect is the Ponce de Leon Hotel, built in St. Augustine, Florida, between 1886 and 1888 (Fig. 57). The architects of the building, Carrère and Hastings of New York, chose poured concrete as the material most likely to withstand hurricane winds. The footings, foundations, main exterior walls, and some interior partitions are monolithic concrete with a shell aggregate. The hotel has been able to resist enormous wind loads partly through sheer mass of material and partly because the monolithic walls form a rigid box. The idea quickly took hold: three other buildings in St. Augustine followed the hotel within a few years, all of them standing today. Ironically enough, however, they were built at the time when plain concrete was being superseded by the reinforced material, which can resist tension and can thus be used for columns, beams, and floor slabs as well as for walls and footings.

Since the arch is subject mainly to compression, it was a logical step to substitute concrete for masonry in arch bridges. The first such structure was a highway span built in 1840 over the Garonne Canal at Grisoles, France. It was thirty years, however, before the concrete bridge was adopted in the United States: the initial essay was a little footbridge constructed in Prospect Park, Brooklyn, New York, in 1871. The span was 31 feet, and the arch had the once universal form of a solid barrel with solid spandrel walls. The concrete vault came in an embryonic form in 1874, when the New York and Erie Railroad lined a rock-cut tunnel at Bergen, New Jersey, with concrete. Most plain concrete bridges that followed the pioneer work were stone-faced structures in which the concrete was limited to the interior volume within piers and abutments and between the deck and the arch barrel. The combination form was adopted for both structural and aesthetic reasons. Since the behavior and the preparation of concrete were still inadequately understood, the builder usually preferred to divide the load between the little-known concrete and the long-familiar masonry. Further, the weathering properties of stone were well known, whereas the absence of uniformity in mixtures of concrete sometimes led to an inexplicable crumbling of the surface after prolonged exposure to the air. As for the aesthetic aspect, stone had the most honorable stylistic associations and until very recently was always superior in texture and color to the synthetic material. The largest of all stone-faced concrete bridges brought the technique to a close: between 1900 and 1902 the Pennsylvania Railroad replaced its iron bridge over the Susquehanna River at Rockville, Pennsylvania, with the present stone and concrete viaduct. This structure has an over-all length of 3,820 feet, divided into forty-eight arches of 70-foot span.

Since plain concrete in mass served well for all elements that functioned mainly through sheer weight of material, it offered many advantages for the structures of waterway control. Indeed, these virtues are so obvious that one can explain its late application to dams and lock walls only on the ground of its unfamiliarity. The concrete dam is a creation of the late nineteenth century, but it came only after a long preparatory development. Dams and dikes in the form of earth or broken-stone embankments date from preclassical antiquity: they were used in connection with irrigation projects undertaken in the early dynastic history of Egyptian and Mesopotamian peoples. Devices for impounding water have had a continuous history since that time, and the basic forms remained essentially unchanged until the nineteenth century. The alternatives were simple and few: earth- or rock-fill embankments, rubble-filled timber cribbing, and fill with an impervious clay core. For small installations, such as those associated with water wheels, the timber weir probably sufficed, but for anything larger the weight of water had to be opposed by the weight of earth or rubble. Immense and spectacular waterway structures of concrete obscure the fact that earth- and rock-fill dams are not only common today but will never be wholly superseded. Indeed, the largest dam in the world—Fort Peck on the Missouri River—and one of the highest in the United States—Oroville on the Feather River—are earth-fill embankments.

But the concrete dam is a necessity for many kinds of installations, and it is the outcome of a progressive development from earth through rubble and coursed masonry to the poured monolith. The fundamental theory of form was primitive until the advent of the masonry dam: the weight of earth or rock had to balance that of the impounded water; the material had to be sufficiently impervious to reduce the passage of water through the dam to a small fraction of the natural flow; the slope of the faces had to be at the angle of respose, that is, the angle at which the earth or rock will not slide down, and gentle enough to prevent excessive erosion. But this rule-of-thumb approach led to wasteful practice in the case of masonry construction. Progress in statics and hydrodynamics eventually taught the builder that the stable form is one with a trapezoidal cross section so calculated that the resultant of the water load and the weight of the masonry mass will pass through the base of the dam within the middle third measured from upstream face (heel) to downstream face (toe). In this way the dam will be prevented from overturning either by tipping backward (upstream) or frontward (downstream). The best calculations, of course, are nullified by inadequate foundations or poor bearing rock, either of which will allow the passage of water under or around the dam, with possibly disastrous consequences.

The early history of dam building in the United States is highly irregular because of the many purposes dams are designed to serve and the variety of resources available to build them. In addition to the familiar needs we now think of, such as flood control, navigation, irrigation, and power, dams have also been built to provide a community water supply, maintain canal reservoirs, and serve a great range of industrial and recreational uses. Most dams built in the colonial and early republican periods were earth- and rock-fill types. The few masonry structures were built up by pouring mortar around loose rubble.

The proper construction of a relatively large earth-fill dam was well exemplified by South Fork Dam, completed in 1839 in the South Fork of the Conemaugh River near Johnstown, Pennsylvania (Fig. 58). The installation was made by the Pennsylvania State Canal, which built twenty-nine dams to impound reservoirs between

Pittsburgh and Harrisburg, an unusual number but typical of summit-level canals in mountainous terrain. Thc author of this well-designed work, William E. Morris, was in no way responsible for the catastrophe that ultimately befell it. South Fork Dam was a combined earth- and rock-fill type with a length at the crest of 840 feet and a maximum height of 72 feet. The upstream half, from face to center, was composed of loose stone facing, earth, and slate core; the downstream half was entirely rubble of large stones. A vaulted sluiceway of dressed masonry in mortar provided a passage for the interior flow of water, which was controlled by valves operated from the top of a masonry tower built at the heel of the dam. The installation functioned very well for its intended purpose, but later owners were careless in their maintenance and allowed the sluiceway to become blocked with sediment. On May 31, 1889, following a two-day rainfall of nearly seven inches, the dam collapsed. The resulting flood, trapped in the narrow valley of the Conemaugh and obstructed by the railroad bridge at Johnstown, nearly obliterated the town and with it about two thousand lives. Since the water could not pass through the sluiceway, it overtopped the dam and quickly swept away the earth fill.

The true masonry dam sprang up very nearly full grown in the United States. The first major work to embody most of the fundamental principles was the original Croton Dam, another of the great masonry structures designed by John B. Jervis for the water supply of New York City. Built between 1837 and 1842 in the Croton River, the dam consisted of a rubble core surrounded by an impervious masonry cover of granite blocks, the whole shaped to the familiar trapezoidal section. Croton was what we now call a gravity dam, that is, one which sustains the water load entirely by the weight of masonry. Old Croton Dam was never demolished: shortly after 1900 it disappeared under the water of the new Croton Reservoir, which was impounded following the completion of Cornell Dam in 1906. The 297-foot height of the present masonry structure far exceeds the 50 feet of its predecessor. Cornell is the highest masonry dam and very nearly the last of the great masonry structures. About 1880 the rubble-core stone-faced dam began to appear in the western canyons to impound water for irrigation and local domestic uses. An early example was the small dam built in the American River near Sacramento, California (1888–89), for the local water supply. Its construction and proportions were typical: rubble core and dressed stone facing; height of 69 feet, with a thickness at the base of 60 feet and at the crest of 25 feet.

As the size of the masonry gravity dam grew rapidly in the West, the system of construction was nearing the end of the road. The structural efficiency could be considerably improved by curving the dam into a vaulted form bowed in the upstream direction, which allows the water load to be transmitted to abutments through the compressive action of the archlike form. The arch dam, however, is best adapted to rock-walled canyons, which provide the best bearing walls for the tremendous thrust of the vault. French engineers of the mid-century, who were the creators of the form, provided the first practical demonstration in Zola Dam, completed in 1843 near Aix-en-Provence. The arch dam appeared in the United States in 1884, when Bear Valley Dam was built near San Bernardino, California, as part of a local irrigation project. The very slender proportions of the cross section indicate that its builders thought of it as a pure arch rather than the combined arch-gravity form that was the rule in the United States until 1963. The arch-gravity principle extended the life of the masonry dam for another two decades, but by that time concrete had carried the day. The last of the masonry works is Roosevelt Dam (1906–11), the first of the big western installations built by the Bureau of Reclamation (see p. 265).

The transformation of the dam from masonry to concrete was a rapid process that seems to have required little more than fifteen years in the United States. Indeed, it is difficult to tell exactly when the process started. Sometime before 1870 the idea had occurred to builders that an inexpensive way to increase both rigidity and impermeability in the masonry dam was to imbed the rubble core in mortar or to build up the core in "lifts" of rubble-and-mortar mix. The initial essay of this kind appears to have been a small dam completed in 1866 at Boyd's Corner, New York. Lynde Brook Dam, built at Worcester, Massachusetts, between 1870 and 1871, carried the process a step farther: the core was formed as a true concrete of cement, sand, and cobble aggregate. In the next decade the Hudson River Water Power and Pulp Company used the technique to build a power dam across the Hudson River at Mechanicsville, New York (1882). The decisive step came four years later when James D. Schuyler designed the original Sweetwater Dam near San Diego, California (Fig. 59). Schuyler's first plan called for a plain concrete arch 50 feet high set in a small, extremely narrow canyon. Construction on this plan began in 1886; the pouring of the arch was largely completed when the builders decided to enlarge it to a height of 85 feet and to use masonry for the additional volume. The resulting structure, completed in 1888, was a composite of concrete and rubble in concrete, with masonry facing on the downstream side and above the concrete mass on the upstream side.

The conditions faced by the engineer in the intermountain deserts and the coastal ranges of the Far West provided him with a special set of opportunities and challenges. The need for water has always been acute, and both the federal government and the southwestern states have had to go to increasingly heroic lengths to satisfy it. With rainfall extremely light and streams far apart, the engineer must build the highest structure he can in order to conserve the maximum quantity of the precious water. But in order to build a high-head dam, he must find a deep, narrow valley walled and floored with rock of the best bearing quality. Finally, in order to impound a reservoir with maximum storage capacity, he must locate the dam where a number of lesser streams enter the main valley at an elevation below the contour level of the dam spillway. It is seldom that all these conditions meet in the ideal geological pattern; and yet, as though to compensate for the limited supply of water, the western mountain regions offer the greatest number of dam sites that approximate the ideal.

The first American dam to embody all the essential features of the modern arch-gravity structure of concrete located in a typical western-canyon site is San Mateo Dam, built in San Mateo Canyon near San Francisco between 1887 and 1889 and enlarged in 1894. The chief engineer of the project was Hermann Schussler and the builder the Spring Valley Water Company, which held a contract for the San Francisco water supply. In both its form and its method of construction, the dam forms the precedent for many of the high-head concrete dams later built by the Bureau of Reclamation. It stands 170 feet high, varies in thickness from 175 feet at the base to 20 feet at the top, and extends 700 feet along the arc. The stream bed was excavated to bedrock by diverting the river through a tunnel in the canyon wall and exposing the bed between rock-fill cofferdams. Concrete was mixed on the site and poured in separate nine-ton blocks with matching grooves and projections; the joints between blocks were then closed by forcing a finer-aggregate concrete into the openings (grouting). Each block had to set from seven to ten days before the adjacent block was poured in order for the concrete to attain reasonable hardness and to dissipate the heat generated by the chemical reactions of setting. The difficult problem of heat generation in an enormous mass of concrete was later solved by circulating cold water in pipes imbedded in the body of the dam. The curved surfaces of the overflow spillway were designed as segments of parabolic cylinders to minimize the turbulence of flowing water, and the toe of the spillway was built out in a long apron which allows the water to lose much of its velocity and thus reduces erosion of the stream bed. The ability of San Mateo Dam to survive the San Francisco earthquake of 1906 with only minor damage enhanced its reputation as a work that helped to usher in the great age of concrete construction.

Before the end of the nineteenth century an entirely new domain of technology was to provide a powerful stimulus to the construction of dams. The generating of electrical power by means of falling water began in 1882, the same year that saw the establishment of the Edison Company's steam generating stations in New York and London. The first hydroelectric plant was installed in the Fox River at Appleton, Wisconsin. A wooden flume carried water to a small timber weir in which valves controlled the flow into a single turbine that was connected by means of a vertical shaft to a generator immediately above it. Hydroelectric generation came to the Far West in 1889 with the establishment of a small plant on the Wilamette River at Oregon City, Oregon. The rock-filled timber crib built to impound the water may be regarded as the original hydroelectric dam. The first masonry structure built for the generation of power was constructed in the

Colorado River at Austin, Texas (1889–93), by the city itself, which thereby introduced the municipally-owned generating system into the United States. The power produced at the plant was used to operate water pumps, electric lights, and streetcars. Austin Dam was a straight-gravity type with an over-all length of 1,150 feet and a height of 65 feet. Its construction was the familiar composite of concrete and rubble core covered by dressed granite facing. It was a bold program, but the life of the dam as originally built was short; it was seriously damaged by flood in 1900 and abandoned as an operating entity. The structure was rescued in 1937 by the Public Works Administration and was greatly enlarged and restored to service in the succeeding three years.

The transformation of the storage dam into a hydroelectric installation immediately extended the range of structural techniques associated with waterway control. The presence of the powerhouse and the passages for the supply and discharge of turbine water added bearing walls, vaulting, and systems of framing in novel forms and arrangements. The main subsidiary structures required for power generation are the powerhouse, control facilities, and switchyard, the last of which is ordinarily not enclosed. Water is conducted to the turbine rotors through penstocks, which are usually imbedded in the body of the dam but may be placed in the rock walls on either side of the stream when the dam is located in a narrow canyon. The penstock intakes are either in the upstream face of the dam or set at one side of the main mass with separate control works. The turbine water flows from the penstock through a sharply curving passage of circular section into the turbine housing and from there to a draft tube, from which it is discharged into the tail race below the powerhouse. As the design of the hydroelectric dam progressed, the passage leading to the turbine was given a spiral form with a diminishing radius in order to accelerate the velocity of flow. The cross-sectional area of the draft tube, on the other hand, expands continuously throughout its length to reduce the velocity of the discharge water and thus minimize the erosion of the stream bed at the tail race. The powerhouse and the control room came to be housed in a single enclosure that grew in size until it reached a volume comparable to that of the largest urban buildings.

It soon became clear that for maximum efficiency of construction and function the entire complex of dam and appurtenant structures would have to be built of concrete. This was accomplished at the very end of the century with the construction of the hydroelectric dam associated with the Grasse River Canal, dug to bypass the rapids of the St. Lawrence River at Massena, New York (1897–1902). The installation was superseded by the enormously larger locks and dams of the St. Lawrence Seaway (1954–59). The great multipurpose dams that now control the entire lengths of major waterways are creations of the present century, but their construction could come about only with the development of new structural techniques and new socioeconomic concepts that began to emerge shortly after the turn of the century.

Chapter Fourteen | Reinforced Concrete: Buildings and Bridges

Because concrete by itself can work only in compression, it will quickly fail if it is used for members subject to high bending forces, such as beams and floor slabs. If it is reinforced with iron or steel bars, however, the elastic metal will take the tensile and shearing stresses, and the rigid concrete will sustain the compressive forces. Exactly why this nice division of labor occurs is not entirely understood, but it can be made exact if the metal is concentrated in the region of maximum tension. As the engineers steadily increased their knowledge of reinforcing techniques, they gradually came to realize that the structural possibilities of concrete are virtually unlimited. In its reinforced form it combines the elastic properties of ferrous metals with its own initial plasticity and ultimate rigidity, an almost paradoxical union of virtues that makes it the most adaptable of all building materials. Indeed, it is the supreme engineering material because it is susceptible of the most exact scientific analysis and can be cast in the most nearly organic form, in which the shape of the structural element most closely approximates the distribution of stresses within it. We can understand why the quantity of concrete in building now exceeds by more than one-third the quantity of all other materials combined, and why it can as readily be cast in a shell less than an inch thick as poured in a monolith with a volume of more than ten million cubic yards.

The history of reinforced concrete in the nineteenth century is more complex than that of iron construction mainly because it was concentrated in a shorter period of time. The idea of reinforcing masonry beams with iron to increase their bending strength was first proposed by French architects in the late eighteenth century, and certain of these experimental members were tested by Emiland Gauthey and Jacques-Germain Soufflot in connection with the building of the Church of Ste. Geneviève (the Panthéon) in Paris. Reinforced masonry spread to England and the United States in the early half of the next century. The possibility of strengthening concrete in a similar way was first realized by William B. Wilkinson of England, who obtained a patent in 1854 for imbedding a grid of wire rope in a concrete slab. From then on inventions flowed at a constantly increasing rate. In the 1860's a number of American patents were issued for reinforced concrete pipe, reinforced joints in brickwork, and timber-reinforced concrete walls. More valuable for structural purposes was Josef Monier's invention in 1867 of a method of reinforcing with wire mesh, which he originally employed as a means of strengthening garden tubs and pots. The decisive step came in the same year, when François Coignet exhibited a technique of reinforcing with bars at the Paris Exposition. Working from Coignet's invention, François Hennebique began to build reinforced concrete floor slabs in 1879 and soon became the leading figure in this newly emerging system of construction. Meanwhile, the methods of experimental science were being applied to these novel techniques, which were wholly without precedent in the structural arts. The chief pioneer was Thaddeus Hyatt, an American by birth who conducted a long series of experiments in London on the behavior of reinforced concrete members. The results were published in 1877 under the title, *An Account of Some Experiments with Portland Cement Concrete, Combined with Iron, as a Building Material,* a work now regarded as the prime source document in the history of reinforced concrete. Hyatt was far ahead of his time: he investigated an extraordinary variety of reinforcing methods, studied most of the physical properties of concrete under all conditions of actual use, and anticipated many basic modern techniques.

By the seventies the stage was set for a fully developed essay in practical building. The author of this valuable achievement was William E. Ward of Port Chester, New York, a wealthy manufacturer who between 1871 and 1876 built his own residence entirely of reinforced concrete (Fig. 60). His architect was Robert Mook of New York, but Ward himself designed the working structure, including the reinforcing techniques. His account of the genesis of the project indicates that he was well abreast of the latest inventions and experiments and that he anticipated some of Hyatt's discoveries. "The incident," he wrote,

> which led the writer to the invention of iron with béton concrete occurred in England in 1867, when his attention was called to the difficulty of some laborers on a quay trying to remove cement from their tools. The adhesion

of the cement to the iron was so firm that the cleavage generally appeared in the cement rather than between the cement and the iron.[1]

Ward began to conduct his own experiments, which eventually led him to a sound understanding of the action of the two materials in conjunction. He concluded that

> the utility of both iron and béton could be greatly increased for building purposes through a properly adjusted combination of their special physical properties, and very much greater efficiency be reached through their combination than could possibly be realized by the exclusive use of either material separately, in the same or in equal quantity.[2]

Ward's house, which still stands, is the first building constructed throughout of reinforced concrete. The floor slabs rest on concrete beams massively reinforced with wrought-iron I-beams ranging in depth from 5 to 7 inches and located in the lower half of the concrete member, where the tensile stress is concentrated. Because the tension increases from zero at the central axis to a maximum at the lower surface, Ward correctly thought that an inverted T-beam would provide a more exact distribution of steel than the common I-section, but the latter had become a standard rolled shape, whereas the T would have had to be specially manufactured. Ward designed the floor slabs with such a generous margin of safety that one would suppose he intended them for bridge loadings. He first laid down a thin slab of concrete directly on the supporting beams; on this he placed a grid of iron rods laid in two courses at right angles; he then poured a 2-inch top slab over the grid and carefully tamped the wet concrete around and under the rods. For a finished floor he added a ½-inch coat of cement and beach sand, placed after the main slab had set. A total of 13,000 square feet of flooring was built up in this way. Ward tested the parlor floor with the fantastic load of twenty-six tons, placed at the center of the 18-foot span. After leaving the weight in place throughout the winter, he found the deflection to be only 0.01 inch.

[1] William E. Ward, "Béton in Combination with Iron as a Building Material," *Transactions of the American Society of Mechanical Engineers,* IV (1883), 388.

[2] *Ibid.*

The coffered ceilings of the house are concrete poured around a grid of I-beams that are arranged in a pattern of squares, each enclosing a slab of concrete laid over reinforcing rods. Plaster was applied directly to the concrete for the fine decorative beading and the finished ceiling. The columns are hollow cylinders reinforced with rods bent into a succession of hoops to sustain peripheral tension—an anticipation of Hyatt's discovery that, since columns are subject to buckling and crushing, the resulting distortions of the shape induce tensile stresses in the region near the outer surface. The most massive structural element in the house is the rear balcony: its eight-thousand-pound weight is cantilevered 4 feet from the wall by means of concrete beams again reinforced with I-beams. All exterior walls and interior partitions are also of reinforced concrete. Ward thought that a properly reinforced 2½-inch slab set vertically would function as the structural equivalent of an 8-inch brick wall, a conclusion that was a prevision of the box frame that was to come sixty years later (see p. 245).

The Port Chester house is a costly tour de force—one would not ordinarily expect, for example, a twenty-six-ton weight on the parlor floor—but it had the inestimable value of a full-scale, highly successful experiment. Ward recommended the use of concrete for the first Hudson River tunnel (1874–1904) and for the base of the Statute of Liberty (1883–86), a recommendation that was followed in both cases; yet he himself never followed up his great achievement, preferring the role of tool manufacturer to that of builder. The development of reinforced concrete for bridges and commercial structures was the work of others, men whose entire careers were devoted to the building arts.

The man who led the way in turning the experimental work of Hyatt and Ward into standard building practice was Ernest L. Ransome. Born in Ipswich, England, he came to the United States shortly after the Civil War, having decided at the outset to settle in California and to embark on a career as a builder. He established a business for the manufacture of concrete block in San Francisco in 1868 and opened a Chicago branch in 1872. For a builder with Ransome's interests, California was a wise choice. San Francisco entered into its boom following the opening of the Union Pacific Railroad in 1869, although multistory commercial building was frustrated by a peculiar set of circumstances. Tall buildings had to be proof against earthquake shocks as well as fire, but the high cost of iron on the Pacific coast made heavy braced construction prohibitively expensive. With unusual farsightedness, Ransome saw that building in concrete offered the best possibility of overcoming the difficulties. On the basis of experimental data that had been established by 1880 as well as his own considerable talents as an inventor, he launched into a career that was to bring him forty years of prosperity and ultimately an international reputation.

Ransome's first essay in reinforced concrete construction was a modest one: in 1883 he built a concrete sidewalk at Stockton, California, which was reinforced with 2-inch rods to take the tension in the unsupported slab. Two years later he built his first reinforced concrete building when he won the contract to construct a flour mill for Starr and Company at Wheatport (an unidentified place, but probably a grain port in the Bay area). Ransome's intention in this design was specifically to secure maximum resistance to earthquake, explosion, and fire. The floor slab was cast integrally with its supporting beams, each of which was reinforced in the tension zone with a single rod (Fig. 61, *top*). Ransome's wine cellar for the Bourn and Wise Company at St. Helena, California (1888), revealed a more advanced form of construction: the floor was poured as a series of parallel sections with arched undersurfaces, the equivalent in concrete of tile arches spanning between iron beams (Fig. 61, *bottom*). The old composite of three materials (iron, tile, and plain concrete) is reduced in the wine cellar to a single slab poured around the paired reinforcing bars in the beams between the arches.

Fully homogeneous floor construction came with the factory of the Pacific Coast Borax Works at Alameda, California (1889), in which the girders, beams, and slab of any one bay were cast as a unit on concrete columns. This was followed by a complete work of reinforced concrete construction, including roof tiles and a domed skylight as well as slabs and columns—the museum built on the campus of Stanford University in 1892. Several industrial buildings in the East, constructed around the turn of the century, rounded out Ransome's career. In the factory of the Varley Duplex Magnet Company in Jersey City (1901) he first used the technique of cantilevering the floor slab beyond the footings to support the wall above, and in the Kelly and Jones factory at Greensburg, Pennsylvania (1903–4), he introduced projecting box girders in the wall plane so that window openings could be continuous and completely independent of the framing system.

Progress in reinforced concrete construction turned the engineers to the possibility of improving the action of plain-concrete arches by adapting the new technique to the building of arch bridges. By the decade of the eighties structural theorists had learned that in semicircular and segmental arches and in any kind under moving, eccentric, and wind loads the pressure line always deviates from the arch axis and that when it does so the arch is subject not only to compression but also to tension resulting from the bending and buckling forces. The traditional practice, as we have noted, was to absorb these tensile stresses in the sheer mass of material, but this redundancy is uneconomical and unscientific. Reinforced concrete offered the possibility of a more efficient design, since the elastic reinforcing could do the work formerly done by the additional volume of concrete.

The highly trained French and German engineers were the first to adapt reinforcing to arch construction, and by the mid-eighties they had built several arch bridges with scientific variations on the methods of reinforcing invented by Josef Monier during the 1870's. In the United States Thomas C. Clarke led the way when he proposed in 1885 to build Washington Bridge in New York as a two-span concrete arch in which the arch barrel was to be reinforced with wrought-iron I-beams inserted at the crown and the haunches. The idea of a 285-foot concrete arch was not likely to be taken seriously in 1885, even when offered by an engineer of Clarke's reputation. The pioneer essay is a more modest work and another of Ransome's contributions to the art—the Alvord Lake Bridge in Golden Gate Park, San Francisco, built in 1889 and still in service (Fig. 62). The solid barrel of the arch is actually a flattened vault, since it spans only 20 feet for a width of 64 feet. The entire structure is concrete throughout, abutmcnts, wing walls, and rails as well as arch, and there is no masonry facing, although the surfaces were carefully scored and roughened to imitate stone blocks. The reinforcing consists of beds of longitudinal bars set near the lower and upper surfaces of the barrel. The system was never duplicated in exactly this form, chiefly because reinforcing with beams soon became the standard and remained so until about 1905, by which date the construction of arch bridges in separate ribs required a more elaborate arrangement of longitudinal, transverse, and radial bars.

The reinforcing of arches was still an unfamiliar technique in the United States during the decade of the nineties, a state of affairs that was reflected in the trial-and-error experimental approach of the builders. Variations on Monier's early and rather primitive system of wire-mesh reinforcing continued to be used for both buildings and bridges past the turn of the century. In the Pine Road bridge over Pennypack Creek, Philadelphia, for example, the arch was reinforced with parallel sheets of mesh distributed in horizontal and vertical planes. More decisive for future progress was the importation into the United States of the efficient system of two-way bar reinforcing developed by the German builder G. A. Wayss from Monier's later patents. The Chicago engineer E. Lee Heidenreich was chiefly responsible for introducing this European technique into America. The Wayss-Monier system was first used in 1899 for storage tanks and grain elevators, but shortly after 1900 it was being incorporated in concrete bridges on the Illinois Central Railroad. Concrete culverts came before the end of the century, and with them came a special kind of reinforcing known as the Luten system, in which the bars were bent into loops conforming to the cross-sectional shape of the culvert.

But the technique most widely adopted was that invented by the Viennese engineer Josef Melan, who secured an American patent on it in 1894. In the Melan system the reinforcing consisted of a number of steel I-beams bent approximately to the shape of the arch axis and laid in a parallel series near the undersurface of the arch (Fig. 63). The resulting structure might be regarded as a combination of the steel-rib arch and the concrete barrel, the concrete serving a protective as much as a structural purpose. The first American bridge to embody the Melan system was a little highway span built at Rock Rapids, Iowa, in 1894. Many Melan bridges were designed by the German-born engineer Fritz von Emperger, who was one of the leading European experimenters in concrete construction around the turn of the century. The most elegant of his spans was the slender, nicely proportioned footbridge built to cross the Housatonic River at Stockbridge, Massachusetts, in 1895. The 100-foot span and 10-foot rise gave the bridge a finely poised, dynamic quality that admirably revealed the architectonic possibilities of the form. The Franklin Bridge in Forest Park, St. Louis (1898), is more conventional in form and hence more closely typical of the Melan spans (Fig. 63).

Emperger was awarded two patents in 1897 for additions to the Melan system, one for placing horizontal I-beams as reinforcing into the deck slab, and the other for inserting radial bars into the spandrel walls to join the arch and deck beams. These innovations involved the use of an excessive quantity of steel, virtually to the point where the bridge became a steel structure with a protective concrete envelope. They may be regarded, however, as steps in the direction of arch-rib and girder spans in concrete. The division of the continuous arch barrel into separate ribs was achieved in the United States by F. W. Patterson, an engineer with the Department of Public Roads of Allegheny County, Pennsylvania. Patterson began in 1898 to design small highway spans in which the deck was supported by two parallel ribs each reinforced with a single curved I-beam. In the same year he introduced the girder bridge of concrete by dropping the arches from Emperger's design of 1897 and pouring the deck structure as a series of shallow arches spanning transversely between longitudinal steel girders that were imbedded in concrete. His final step came close to the true deck-girder bridge in concrete: he abandoned the transverse arches, reduced the steel girders to two, on at each side of the deck, and poured concrete around them. The result was a bridge in which most of the deck load was taken by the steel members, but it marked the beginning of true concrete-girder design.

One other innovation served to round out the development of construction up to the end of the century. In 1894 David A. Molitor proposed a three-hinged arch of concrete for a bridge with a single span of 236 feet. The hinges were to be pinned connections of steel, and the ends of the concrete arch segments were to rest in rectangular steel sockets fixed to the abutments. The first arch of this kind was built at Mansfield, Ohio, between 1903 and 1904. The form has been widely adopted in Europe, especially by French and Italian engineers, but is seldom used in the United States. The Mansfield bridge serves to illustrate the contrast between the experimental and enterprising spirit characterizing American concrete construction around 1900 and the conservative attitude of American builders compared to European that emerged in the later decades of the century.

Part Four

Industrial and Urban Expansion in the Twentieth Century

Chapter Fifteen | The Steel Frame: Skyscrapers and Railroad Terminals

Since the evolution of building techniques is a continuous process, the establishment of 1900 as a dividing point is bound to be somewhat arbitrary. The major inventions of the twentieth century, especially those in concrete construction, did not play an important economic role in American building until the end of World War II. Up to this time the framing systems of tall buildings, the main structural features of truss, suspension, and arch bridges, and the structures of waterway control represented progressive developments of techniques that had been created and brought close to maturity in the previous century. At the same time, the accelerating economic expansion of the country and the shifting patterns of industry and transportation brought about profound changes in the material fabric of the city. As a matter of fact, although the man-made environment as we know it today—the high curtain-walled buildings, suburban growth, the proliferation of structures for air and automotive transportation, multipurpose systems of waterway control—was largely built after World War I, the fundamental structural character of these creations, in most cases, was established in the first decade of the century. New York was easily in the front rank at the time: the first group of East River bridges after the Brooklyn span was opened before 1910; the two great railroad terminals were started and largely completed in the same period; the skyscraper had passed thirty stories in height and shortly after 1910 was to reach the fifty-five stories of the Woolworth Building. At the other end of the geographical scale, so to speak, the uninhabited deserts of the West were to be radically transformed by the establishment in 1902 of the Bureau of Reclamation. Railroad truss and girder bridges were built at a steady rate year after year until 1930, when the century of rail dominance came to an end. The basic truss forms had been substantially fixed by 1900, with the exception of the continuous truss, but that too was a product of the opening decade.

The riveted and wind-braced steel frame was the creation of the Chicago builders in the period of 1880–1900, but the largest projects to embody these structural principles were concentrated in New York at the turn of the century. Chief of these are the two Manhattan railroad stations, unique engineering works of such size, cost, and

64. Pennsylvania Station, New York City, 1903–10. McKim, Mead and White, architects; Charles W. Raymond, chief engineer. Interior of the main concourse. The two New York stations, which included extensive tunnel construction and electrical installations as well as steel and concrete structures, constituted the greatest works of building art undertaken in the United States by means of private investment.

65. Grand Central Terminal, New York City, 1903–13. Reed and Stem, Warren and Wetmore, architects; William J. Wilgus, chief engineer. Cut-away drawing showing the concourse and track levels. In spite of the size and complexity of station and track facilities, the terminal is a masterpiece of simple, functional planning.

66. Grand Central Terminal. The track area during construction.

67. Woolworth Building, New York City, 1911–13. Cass Gilbert, architect; Gunvald Aus Co., engineers. Elevations and details of the steel frame. The Woolworth was the highest building in the world at the time of its construction and was supported on the most elaborately braced steel frame.

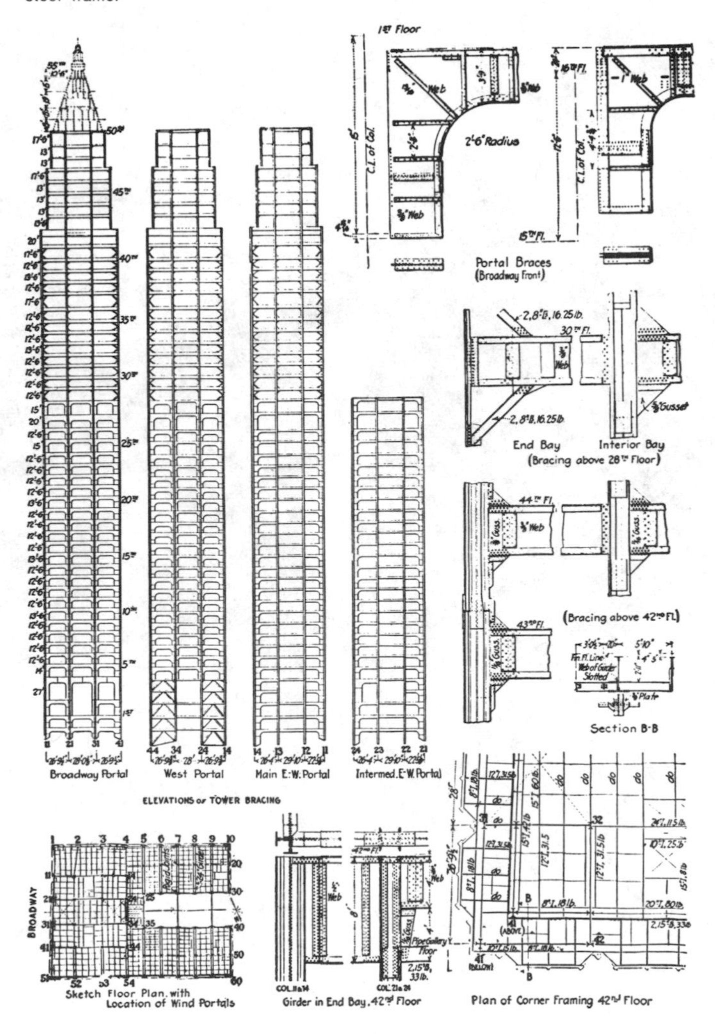

68. Empire State Building, New York City, 1929–31. Shreve, Lamb and Harmon, architects; H. G. Balcom, engineer. Elevation of the steel frame. The highest of all American skyscrapers until the World Trade Center is completed, the Empire State is carried on a traditional portal-braced steel frame.

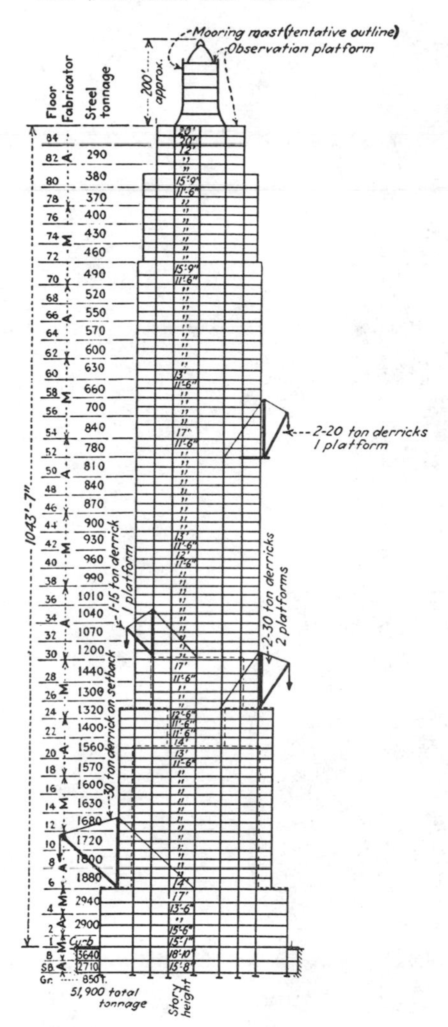

69. Civic Center, Chicago, Ill., 1963–65. C. F. Murphy Associates; Skidmore, Owings and Merrill; Loebl, Schlossman and Bennett: architects and engineers. The structural system of the Center is the largest welded steel frame so far erected, and it includes the longest bay span in a multistory building.

70. Metropolitan Fair and Exposition Building, Chicago, Ill., 1958–60. Shaw, Metz and Dolio, architects; Victor Hofer and Ralph H. Burke, Inc., engineers. Interior view during construction, showing the rigid frames supporting the roof. The aim of extensive open space is best realized in Chicago's exposition building, where only eighteen frames carry a roof with an area of 357,000 square feet.

71. Beinecke Rare Book and Manuscript Library, Yale University, New Haven, Conn., 1962–63. Skidmore, Owings and Merrill, architects and engineers. In spite of the rich granite and marble dress, the main enclosure of the library is carried by welded steel Vierendeel trusses.

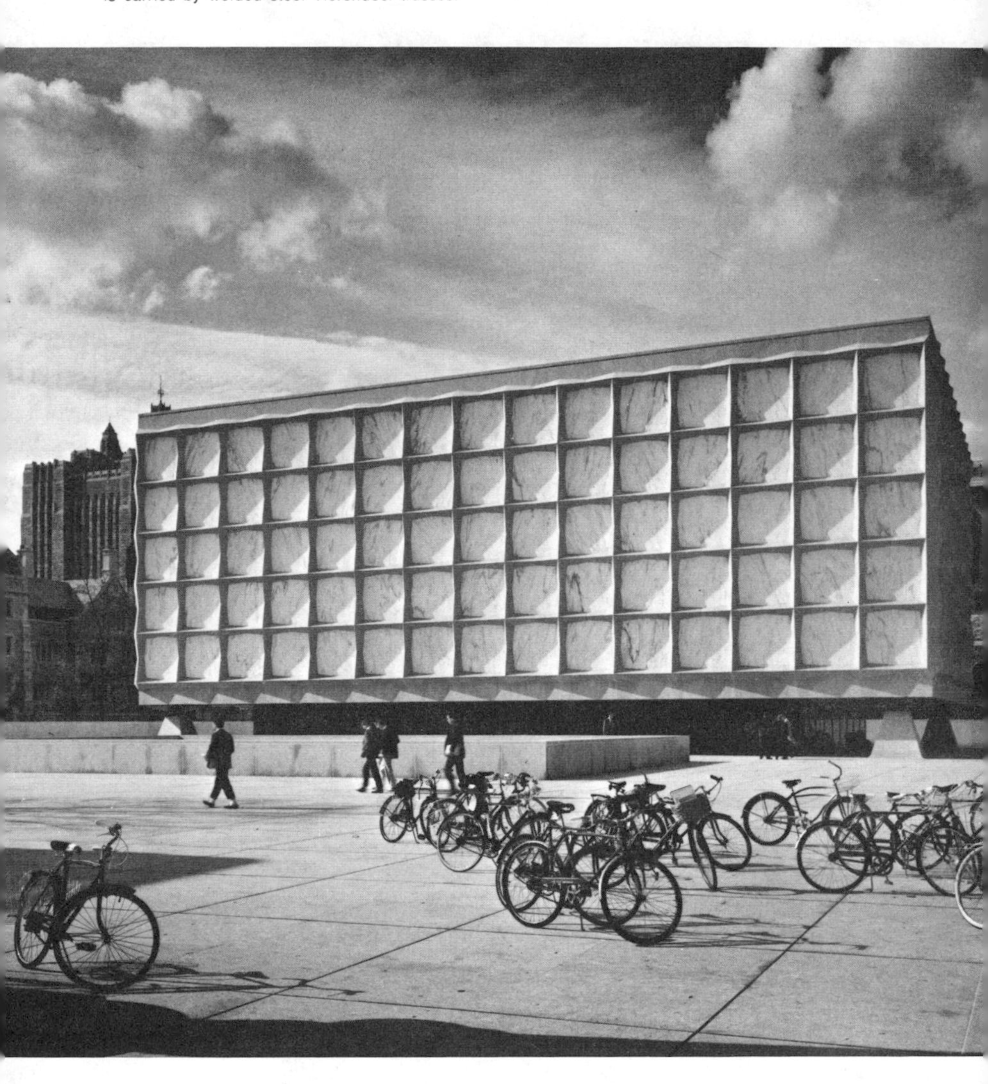

72. IBM Building, Pittsburgh, Pa., 1962–63. Curtis and Davis, architects; Worthington, Skilling, Helle and Jackson, engineers. The bearing walls of tho IBM Building represent a revival of the lattice truss in welded steel.

73. Harris County Stadium, Houston, Tex., 1962–64. Lloyd and Morgan, Wilson, Morris, Crain and Anderson, architects; Praeger, Kavanaugh and Waterbury, engineers. Roof framing during construction. The largest of all domes, the roof of the Houston stadium is supported by a system of skew and radial trusses.

74. Travel and Transport Building, Century of Progress Exposition, Chicago, Ill., 1933–34. Bennett, Burnham and Holabird, architects; B. M. Thorud and Leon S. Moissieff, engineers. The first suspended structure among American buildings.

75. Bridge of the Chicago, Burlington and Quincy Railroad, Ohio River, Metropolis, Ill., 1914–17. Ralph Modjeski and C. H. Cartlidge, engineers. The channel span of the Metropolis bridge is still the longest simple-truss span ever built.

76. Bridge of the Chesapeake and Ohio Railway, Ohio River, Sciotoville, Ohio, 1914–17. Gustav Lindenthal, chief engineer. The continuous truss is best exemplified by the symmetry, clarity, and power of this huge bridge structure.

75

76

77. Bridge of the Texas and New Orleans Railroad, Pecos River, Langtry, Tex., 1942–44. T. and N. O. Engineering Department, engineers. The most elegant of all railroad truss bridges replaced the high cantilever span illustrated in Fig. 53.

78. Pulaski Skyway, Jersey City, N. J., 1930–32. Sigvald Johannesson, engineer. The cantilever truss at last appeared in a graceful form in the Hackensack and Passaic River bridges of Pulaski Skyway.

79. Highway bridge, Atchafalaya River, Morgan City, La., 1931–33. N. E. Lant, engineer. K-truss of the channel span.

78

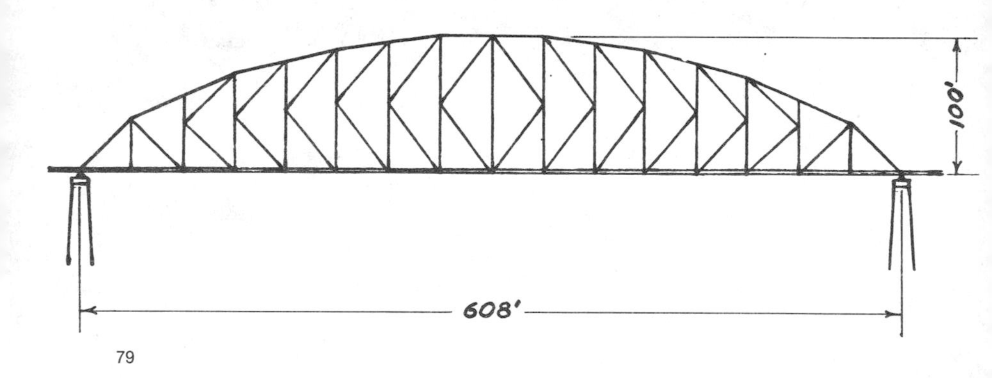

79

80. Central Avenue bridge, Arroyo Verdugo, Glendale, Calif., 1937–38. L. T. Evans, engineer. Vierendeel truss span.

81. Hell Gate Bridge, New York Connecting Railroad, New York City, 1914–16. Gustav Lindenthal, chief engineer. In its over-all size, mass, and dignity Hell Gate is still the foremost steel arch bridge. The little company that built it connects the Pennsylvania and New Haven systems in New York and is owned by the two larger roads.

80

81

82. Highway bridge, Kill van Kull, Bayonne, N. J.–Staten Island, N. Y., 1928–31. Othmar H. Ammann, chief engineer. Once the longest steel arch in the world, it unites the Staten Island expressway with the streets of Bayonne, N. J., and the surrounding communities.

83. Manhattan Bridge, East River, New York City, 1901–9. O. F. Nichols, chief engineer. The Manhattan span typified the post-Roebling suspension bridge, in which a steel framework took the place of masonry in the towers and deep stiffening trusses provided the necessary aerodynamic stability.

82

83

complexity as to be without parallel among the structures financed by private capital. The planning of Pennsylvania Station anticipated that of Grand Central by several years, and its opening came two years before that of its neighbor on Forty-second Street. The concept and its triumphant execution put the Pennsylvania's project squarely in the great tradition of Roebling and Eads—American building at its best—and the demolition of the station can only be regarded as a national calamity (Fig. 64). The architects of the building were McKim, Mead and White, and the chief engineer of the whole system of tunnels, tracks, and structures was Charles W. Raymond, but the man who fathered the original idea was the company's president, Alexander Cassatt. The story of how this great program was conceived and sketched out is clearly apocryphal, but like all legends it contains an element of truth about the personalities involved. Cassatt and Charles Follen McKim were said to have created the whole project in an afternoon's conversation over a bottle of Madeira at Delmonico's Restaurant. They were big men who planned on the heroic scale, and the Pennsylvania Railroad commanded the means to execute their vision.

By 1900 the need for a New York station had become urgent. The Pennsylvania's terminal for the metropolitan area, completed as recently as 1892, was located in Jersey City, from which passengers had to take the ferry across the Hudson to reach the city or the connecting trains to New England and eastern Canada. The situation on the other side of Manhattan was much worse: the Long Island Railroad terminated at Queens and Brooklyn, leaving the many thousands of commuting passengers dependent on the East River ferries and the single car line on Brooklyn Bridge. The other East River bridges and the city's extensive subway system still lay in the future, although they were to come within the next ten years. There had been a number of proposals for bringing the Pennsylvania's tracks into Manhattan and effecting a junction with the Long Island, the earliest of them dating from 1892, but the company was discouraged from acting on them by the seemingly insurmountable obstacle presented by two broad tidal rivers carrying an enormous waterborne traffic. By 1900, however, two events had occurred that pushed the railroad's officers closer to their ultimate decision. One was the extension of the Orleans Railway into Paris through an electrified tunnel, with the associated construction of the electrically operated Gare d'Orsay, the company's Paris terminal (1897–1900). The other was the promise of completion of the first Hudson River tunnel (1874–1904) after twenty-five years of mismanagement, bankruptcy, and fatal accidents.

The Paris tunnel constituted the precedent for Cassatt's own plan, but it was the experience of the builders who finally learned how to master the difficulties involved in completing the Hudson tunnel that provided essential guidance to the railroad's engineers. In 1901 Cassatt decided on the straightforward solution, staggering in its boldness and its scale: he proposed a direct connection between the Pennsylvania's line at Newark, New Jersey, and the Long Island at Sunnyside in Queens, all of it to be underground from the west face of Bergen Hill in New Jersey to the west end of the projected coach yard in Sunnyside. The program included, from west to east, the following major parts: a cut-off over the New Jersey meadow from the Newark station to Bergen Hill; two single-track tunnels under the hill, the Hudson River, and the West Side of Manhattan Island; a twenty-one-track station facing Seventh Avenue between Thirty-first and Thirty-third streets; and two double-track tunnels under the East Side and the East River. The total length of new line would be 13.4 miles, of which 1.5 miles were to lie in subaqueous tunnels. The extensive underground trackage required the electrical operations of trains, a technique that was only six years old in the United States at the time that Cassatt made his decision. The great size and intricacy of the project induced Raymond to organize construction as four separate enterprises corresponding to the four distinct parts of the plan. These were designated in west-to-east order as the Meadows, North River, Terminal, and East River divisions; work on the four was carried out simultaneously, each with its separate group of engineers and laborers. Construction began on June 10, 1903, and was completed for public use on November 27, 1910.

The geological complexity of New York harbor required the

employment of three radically different methods of tunneling. The bed of the Hudson is a mixture of soft alluvial muds and marine sediments, in which the tunnel could be mined by forcing an iron shield through the mud; the mud in the path passed through several openings in the shield and was carried away on rail cars. This technique was invented by James H. Greathead for the first London subway (1887–90) and adopted by Charles M. Jacobs and John V. Davies, the British engineers who finally completed the original Hudson tunnel. If the materials to be excavated are soft enough, as they are in the bed of the Hudson, the Greathead shield eliminates the extremely hazardous practice of mining in front of the shield. On Manhattan Island the technique had to be shifted drastically: here the workers drilled and blasted their way through the hard schist and dolomite of the island's spine. The sand and gravel bed of the East River offered still another kind of problem. Since the material is not fluid like saturated mud, the shield could not be forced through the bed; instead, the way had to be cleared in front of it by maintaining caisson pressure while the workers cut their way into the bed through openings in the shield. But there was worse danger than this naked exposure: the porosity of the material resulted in frequent blowouts, in which the compressed air behind the shield burst through the river bed, with a consequent loss of pressure and an inrush of sand and water. The builders again turned to a technique developed by Jacobs and Davies for the Hudson tunnel. Known as the clay-blanket method of tunnel mining, it consists essentially of blanketing the entire river bed along the line of excavation with impervious clay to prevent the air from blowing through the bed into the water above—a heroic operation that required the placing of 300,000 cubic yards of clay. As the cutting shields progressed through the river beds, the tunnels were built up of successive cast-iron rings bolted together through circumferential flanges. With the iron in place, the whole tube was then lined with concrete and an inspection bench of concrete poured along one side. The tunnel clearance is small enough that the piston action of trains ventilates the tubes.

The narrow bores end approximately on the lines of Fifth and Tenth avenues, between which the right of way expands into the 28-acre area of the station track layout. The entire area was excavated to a maximum depth of 50 feet, which required the removal of 2,441,000 cubic yards of earth and rock. The track level of the station proper includes sixteen through and five stub-end tracks served by eleven island platforms of reinforced concrete (part of the track and platform system appears in the old view shown in Fig. 64). The station building was a riveted steel-framed structure comprised of four distinct structural systems: massive framing of columns and girders to carry the main concourse over the track area below it; similar framing of less weight for the shorter spans in the smaller enclosures, such as waiting rooms, restaurant, stores, and subsidiary passages; conventional truss framing over enclosures with intermediate spans, such as the taxi rotunda; and arch-rib and arched-truss framing to support the vaulted glass and steel skylights over the main concourse. This last structure constituted the most impressive spatial and technical feature of the station building and the only one in which the underlying steel construction was frankly presented to view. Elsewhere the architects hid the steelwork under the envelope of plaster or stone carefully designed to simulate the masonry construction of large Roman vaulted buildings like the Thermae of Caracalla, which formed the architectural precedent for the station building. In the concourse structure the primary bearing members were long built-up columns of steel angles and strap rising from the track level to the springing points of the main vaults. The weight of the glass envelope was sustained by a rectangular grid of steel arch ribs that were carried in turn by the arched trusses, which served to transmit the skylight load to the columns. The ancestors of this brilliant tour de force in steel-arch framing were the glass and iron roofs that covered the interior light courts in the office and hotel buildings of the eighties and nineties. In Pennsylvania Station, McKim, Mead and White carried the idea to its ultimate expression in American building, but electric lighting and a miserly distaste for monumental spaces have relegated it wholly to the past.

Few changes were made in the station during its life of better than half a century. The track level was later entirely covered and an incoming concourse was built on the level immediately below the main concourse. The original system of third-rail direct-current operation of trains was replaced in 1933 by an 11,000-volt alternating current installation with overhead wires. On the completion of Hell Gate Bridge in 1916 (see pp. 229–31), the Pennsylvania's terminal became a through station for trains running from Washington and Philadelphia to Boston and Montreal. At the time the facilities were opened in 1910 the railroad had invested $112,000,000 in what it modestly called its New York Extension, but the current replacement value would be around $700,000,000. Declining rail traffic, rising property taxes, and mounting costs of maintenance were enough to persuade the railroad company to abandon the great building in favor of a profitable air-rights development on the site. The structure was demolished between 1963 and 1966 to make way for a trivial complex of hotel, sports, and office buildings under which the tracks and other facilities are buried.

The construction of Grand Central Terminal began on June 19, 1903, nine days after the Pennsylvania project was started, but the difficulties of maintaining one terminal while another was built in its place added twenty-six months to the time of completion. The station was opened to the public on February 2, 1913, and the full system of loop tracks was placed in operation in 1927. Grand Central did not involve the extensive tunnel work of Pennsylvania Station, but its immense size and its unique plan place it at the same level of building art. Nor did the Vanderbilts' decision to build the new terminal emerge from ten years of reflection and preliminary planning; indeed, it seems to have been made quite suddenly in the light of unexpected exigencies. At the turn of the century the Pennsylvania's intention to construct an electrified station in Manhattan was well known at least in railroad circles. The bitter competition between the two roads might have been enough to move the Vanderbilts to take similar action, but their hands were forced partly by events beyond their control. In January, 1902, a rear-end collision in the smoke-filled Park Avenue tunnel took seventeen lives. The state legislature at Albany responded to the public outcry by passing a law requiring the electrical operation of all trains on Manhattan Island after 1910, although the electrification of the existing Grand Central Station had been inaugurated in 1906. As a consequence, in a moment of reckless inspiration the officers of the New York Central companies decided to replace the whole Grand Central complex with a new terminal.

The author of the grand scheme as it finally took shape was William J. Wilgus, chief engineer of the New York Central and Hudson River Railroad. The most valuable feature of his proposal for the urban economy was the idea of rebuilding the twelve-block area cleared for the track layout by constructing the buildings directly over the tracks. The result was the transformation of an urban desert into valuable property where hotel and office buildings form an integral part of the transportation complex. The heart of the Wilgus plan is the station building facing into Park Avenue at Forty-second Street (Figure 65). The blocking of the avenue by the old building was eliminated by carrying the artery around the new station as a pair of elevated drives. The baggage and express terminal extends north of the central structure to Forty-fifth Street, and the mail terminal lies on the east side along Lexington Avenue. The width of the track layout is two and a half times that of the station building, the extreme size of the track area arising from the need to handle the nation's largest volume of both through and suburban traffic in a single stub-end terminal. The tracks are divided between two levels, the lower of twenty-seven tracks ordinarily reserved for suburban trains, and the upper of forty-three tracks for through trains (both figures include storage and loop tracks). Even with this unparalleled area of trackage, the problem of moving empty trains in and out of the terminal during periods of peak traffic led to such complications that the original layout had to be supplemented by a double set of loop tracks; the upper level was added in 1917, and the lower in 1927.

Construction of the new terminal was carried out by progressive demolition of the old facilities and their replacement with the new. The great balloon shed of the original station, the marvel of another age, was the first to go. Out of the chaos of excavation, demolition, and track laying rose the most elaborate system of steel framing ever built. The structural problem was fantastically complex: first was the frame of the station building and its appurtenant structures; around these spread the massive bridge-like structure of the upper-level track floor; above this lay the viaducts that carried the streets;

and finally there were the independent framing systems of the air-rights building (Fig. 66). All the techniques of riveted steel framing developed up to the time went into the realization of Wilgus's dream.

The structural system of the station and subsidiary buildings is a conventional frame of columns, girders, and floor beams up to the vault over the main concourse and the roofs and skylights above the buildings. The vault is carried on steel ribs hung from trusses that span the width of this magnificent room; the gable roofs and skylights are carried on triangular trusses. Elaborate precautions were taken to prevent damage to columns and bracing from derailed trains and from the vibrations of moving trains, including concrete envelopes around steel members, concrete walls between columns, and special shock-absorbent footings. The steelwork of the station group is covered with granite, marble, and limestone sheathing, costly materials appropriate to the French Baroque architecture of the building, which was initially designed by Reed and Stem and completed by Warren and Wetmore. Among structures outside the building proper, the frame of the express track floor has the massive density common to railroad bridges. The primary members are columns and girders, the latter reaching a maximum of 10 feet in depth; the beams spanning between the girders carry a reinforced concrete floor, 8 to 10 feet thick, cast on the undersurface as a series of little vaults spanning between the beams. This floor structure was designed for a maximum combined dead and live load of 1,815 pounds per square foot. The overhead viaduct structure is a much lighter column-and-girder system designed for street rather than rail loads.

It seems incredible that the operation of trains can be carried on conveniently and safely in this dark forest of steel and concrete; yet the enormous volume of traffic has been handled day after day for more than half a century without a serious accident. The great capacity of the terminal, up to one thousand train movements a day, arises from the four basic operating characteristics, which together constitute a unique system: the double-level track layout, with separation of through and suburban traffic; the two-level loop, which makes it possible to turn trains around without a reverse movement; the electrical operation of trains; and the movement of trains in both directions on all four approach tracks, an innovation of the Grand Central plan that requires an expensive system of double-direction signaling and frequent crossovers from one track to the next.

All air-rights buildings constructed over the track area after the completion of the terminal are supported on columns independent of the track-floor and viaduct framing. The tracks lay open between the streets north of the station for nearly ten years, war and post-war depression preventing the realization of the air-rights program until 1922. The first building to go up was the Park-Lexington (1922–23), and its structural engineer, H. G. Balcom, was the first to solve the multitude of problems that this novel kind of construction posed. It was the beginning of a brilliant career for Balcom, who had the structural systems of the Empire State Building and Rockefeller Center to his credit before he retired from the profession. With the Park-Lexington to lead the way, the air-rights buildings rose in an unbroken series until, with the completion of the New York General Building in 1932, the site was covered. The enormous building boom that began in New York in 1950, however, resulted in the demolition of many of the earlier works and their replacement by the fashionable curtain walls of glass and enameled steel. The latest and largest of these, erected over the north end of the terminal group, is the Pan American Building (1961–63), an exhibition of structural ingenuity but indifferent in its architectural design and reared with reckless disregard for the principles that underlay the primary vision. But the New York Central and the New Haven railroads have a desperate stake in these rents, for the larger company frequently loses money on its rail operations, and the smaller is bankrupt.

At the time the New York stations were completed the skyscrapers were still concentrated in the downtown area around Wall Street and lower Broadway. There was none among them that matched the constructive genius lavished by the railroad companies on their midtown facilities until the Woolworth Building was erected at 230 Broadway between 1911 and 1913 (Fig. 67). Cass Gilbert was the architect, and the Gunvald Aus Company were the structural engineers. It was the highest building in the world at the time, rising through fifty-five stories to a total height of 760 feet 6 inches at the top of of its pyramid roof. The vivid screen-like walls with their delicate and aerial Gothic ornament perfectly express this upward-soaring power but tell us little of the heavily braced steel frame that supports the multitude of floors and keeps the tower rigid against the turbulent winds of New York. The wind-bracing of the riveted frame is for the most part a system of portal arches like those of the Old Colony Building in Chicago (see p. 129). The arched frame of the Woolworth tower extends up to the twenty-eighth floor; above this, to the forty-second floor, bracing is secured through a double system of knee braces in which the knees are located on the top and bottom of the girder diagonally across the corners at the connections between the girder and its supporting columns. The Woolworth frame could easily withstand hurricanes of maximum intensity; however, this lavish distribution of steel in deep fillets and braces came to be regarded as an expensive redundancy of metal with an unnecessary sacrifice of the vertical space between the floors. The development of high-strength steel, welded connections, and new techniques of riveting and bolting eventually made it possible to eliminate these additional shapes in buildings even higher than the Woolworth.

Progress in riveted framing was a kind of orthogenesis in which redundancies were refined away as the height of the New York skyscraper climbed steadily upward. The record among completed works, which will be exceeded by the twin towers of the World Trade Center (see pp. 197–98), is held by the Empire State Building, constructed between 1929 and 1931 from the plans of the architects Shreve, Lamb and Harmon and the engineer H. G. Balcom (Fig. 68). The main shaft of the building, which embraces the rentable floor space above the setbacks, rises eighty-five stories to a height of 1,044 feet above grade. A squat pylon, once thought to be useful as a mooring mast for airships, extends upward another seventeen stories, for an over-all height of 1,239 feet. The supporting structure is a standard riveted steel frame with the simplest kind of portal bracing, in which the girders are riveted throughout their depth to the column web and the beams riveted in the same way to the girders. The most impressive feature of the frame was the speed with which the steelwork was erected: by means of derricks located on the numerous setbacks, the contractor put 51,900 tons in place during six months of 1930. The familiar setbacks, required by the New York Zoning Law of 1916, introducd two deviations from straightforward framing at the sixth and seventy-second floors, where offset columns bear on specially enlarged members called distributing girders. Skyscrapers of New York constructed since the Empire State have revealed no marked departure from the conventional frame except for the introduction of welding and the revival of the older technique of bolted rather than riveted connections.

The Chicago building boom of the twenties reached its full momentum about 1928, a few years behind the boom of New York. The skyscrapers of the midwestern city have been uniformly lower in height, seldom exceeding forty stories for office towers, but exigencies arising from peculiar features of location and design have led to special variations on the usual framing techniques. The best-known and most controversial of the Chicago skyscrapers is the Tribune Building, which represents one of the rare cases where complexities of architectural design wholly unrelated to modern construction demanded the most ingenious structural solutions. The Tribune was built between 1923 and 1925, following an international design competition won by Raymond Hood and John Mead Howells. Henry J. Burt was associated with them as structural engineer. The building is located on the north approach of the Michigan Avenue bridge, where the street is elevated above grade level to provide clearance for the double-deck span over the Chicago River. The thirty-four stories extend 450 feet above the upper level of the street, and the seven sub-basements for the printing plant and newsprint storage extend 41 feet below the water level in the river. The presence of ground water in the soil of Chicago has always made pumping necessary to maintain dry excavations for deep foundation work, but in the case of the Tribune this reached heroic proportions, for continuous pumping against high-pressure seepage had to be combined with braced cofferdams of sheet-steel piling.

The frame of the building is noteworthy for the number of offset columns, especially at the twenty-fifth floor, where the main shaft gives way to the octagonal tower, and at the thirty-third and thirty-fourth floors, where the diameter of the tower is sharply reduced. All column lines break at the twenty-fifth floor, so that the entire framing system of the tower has to be carried on a dense grillage of girders in the twenty-fifth floor plane and on a pair of 43-foot long trusses extending in depth from the twenty-sixth to the twenty-seventh floor. The eight "flying buttresses" around the tower are supported by hollow built-up columns each in the form of a slender octagonal cage joined to the tower frame by a single horizontal strut near the top of the column. Needless to say, none of this steelwork arises from structural necessity: the false buttresses were designed to provide a proper visual transition in the Gothic style from the main shaft to the tower. The wind-bracing of the Tribune was designed with special care because of the slender tower-like form and the exposure to the turbulent winds of the lakeshore. Knee braces above and below the girders are the chief bracing elements, but these are supplemented by full-bay diagonal bracing in the east and west sides of the elevator shafts and by full-bay horizontal diagonals in certain of the lower floors.

The completion of Chicago Union Station in 1925 paved the way for the first air-rights development in the city. The Daily News (now Riverside Plaza) Building, designed by Holabird and Root and the engineer Frank E. Brown, was erected between 1928 and 1929 over the area where the north approach tracks diverge into the fourteen station tracks. The column spacing under the eastward-projecting wings and the plaza was thus determined by the irregular pattern of switches and tracks at the basement level. Under the north wing, for example, the column spacing in places exceeds 100 feet, with the result that the frame above the grade level had to be hung from trusses located under the roof of the wing. Under the south wing, long spans and irregular clearances required a complex pattern of trusses and cantilevered girders: part of the frame is suspended from a 57-foot truss under the roof, and the rest is supported by a 91-foot girder. The west end of this member is extended as a cantilever to carry a line of columns for the twenty-six-story office building at a point 18 feet 6 inches from its support, an offset load that required a girder 10 feet deep.

The central tendency in steel-framed construction has been to avoid members with extraordinary dimensions and shapes, to simplify connections, to standardize wherever possible, and to reduce the total number of framing members, especially columns. The fullest realization of this program depends on technical innovations such as welding and rigid framing and the application of rigid-frame trusses to buildings, which we will consider in their appropriate places (see pp. 192–98). The ideal framing system for an office building is one in which there are no columns within the walls of the building, and the ideal plan to accompany it is one that separates utilities and service facilities from the rentable area of the office floors. There are a number of structural and planning means by which these ends can be achieved, but the initial steps were taken with conventional framing techniques. The first office block with a column-free interior is the Gulf Oil Building in Pittsburgh, constructed from 1931 to 1932 from the designs of the architects Trowbridge and Livingston and the engineers Weiskopf and Pickworth. The structure was planned as a hollow square with all columns restricted to the wall planes and the periphery of the court. The open core is given over to elevator shafts and other utilities, and the surrounding floor areas are reserved for office space. This hollow-square plan, as we have seen, was a common feature of office and hotel buildings of the late nineteenth century, but in the earlier works the purpose was to provide light for inner offices. In the Gulf Oil Building, however, the aim was to provide an uninterrupted work area on every floor, which could be realized through wider bay spans and progress in artificial lighting. Since the court is a rectangular area enclosed within a square, the distance across the intervening floor area varies. The maximum bay span is 35 feet 11 inches, not a remarkable length but considerably longer than the standard of the time for conventional framing systems.

Further progress in the hollow-core plan came with the Continental Center in Chicago, designed by C. F. Murphy Associates and built from 1961 to 1962. The column-free interior is maintained by adopting square bays with an unprecedented uniform span of 42 feet, which require primary floor girders two feet deep. The size of the girders entailed a considerable sacrifice of vertical space for the floor framing, but surprisingly enough, other economies emerged to offset the higher cost of the exaggerated floor-to-floor height. By cutting rectangular openings into the girders, the engineers transformed them into rigid-frame trusses through which air ducts, pipes, and electrical conduits could bc passed rather than suspended from their undersurfaces. Further, the reduction in the number of columns correspondingly reduced the number of footings and caissons, which resulted in a saving in the cost of foundation work. The design of the Continental Center marks a vigorous return to the architectural principles of organic design and structural expression as these were developed by the original Chicago school: the wide-bayed bolted frame is directly expressed by the reduction of the street elevation to a rectangular pattern of steel plates that cover the fireproof cladding of the outer columns and the spandrel girders.

Chapter Sixteen | New Departures in Steel Framing

Technical innovations are ordinarily structural forms with their appropriate modes of assembly, but one of the most useful inventions of the twentieth century has to do exclusively with the way the members of a structure are united. Electric arc-welding offers the great advantage of fusing two pieces into a solid homogeneous unit, which obviously provides a stronger joint than one that connects members by means of additional shapes and numerous perforations for the rivets or bolts that bind them together. But for many years after its initial appearance, welding possessed two serious defects, namely, that external inspection could not reveal whether the weld produced solid fusion over the contact areas, and that extreme differences of temperature in small regions produced dangerous secondary stresses when locked permanently into the metal. A long period of experimenting with materials, manual techniques, and radiographic analysis was necessary before welding could become a reliable method of joining members. The original form of arc-welding was invented in 1881 by Auguste de Meritens, one of the French pioneers in the development of the electric generator, but it was not applied to building until after the turn of the century. It was first employed in the United States to join the separate elements of wire-mesh reinforcing for concrete. This minor role in the whole construction project showed little promise for future development, and more than a decade passed before a further step was taken.

The initial application of welding to a building frame appeared in the factory of the Electric Welding Company of America, built in Brooklyn, New York, in 1920. The structural engineer, T. Leonard McBean, specified it for the joints in the Pratt trusses that supported the gable roof but retained the conventional riveted connections for the rest of the frame. The electrical industry was understandably interested in the new process, and it was chiefly through the impetus provided by the Westinghouse Electric and Manufacturing Company that welding reached its maturity in American building. The total break with the older techniques of joining steel members came in 1926, when the Westinghouse company constructed its factory at Sharon, Pennsylvania. The frame of this five-story building was specifically designed by the company engineers as a complete welded structure, a rigid and homogeneous grid of interconnected columns and beams. By 1932 Westinghouse had erected twenty-six buildings with welded steel frames, including the eleven-story Central Engineering Laboratory in Pittsburgh (1930). The Westinghouse experience provided a convincing demonstration of the reliability of the technique, and in the 1930's it was widely adopted for industrial buildings, especially the new one-story glass-walled factories of the automotive and aircraft industries. It was particularly useful in building up continuous girders into shapes that would match the saw-tooth and trapezoidal profiles of roofs with projecting light monitors. The elegant framing systems that resulted offered many advantages over the truss, with its multiplicity of light-obstructing members.

The long hiatus in building caused by the depression of the thirties and the war that followed drastically limited the application of welding to high-rise construction until the boom that began in 1950. The full potentiality of the technique came with the Inland Steel Building in Chicago, erected between 1955 and 1957 from the plans of Skidmore, Owings and Merrill. In this brilliant exhibition of structural virtuosity, all floor and roof loads are carried by 58-foot transverse girders to seven pairs of columns located outside the wall planes on the long elevations. As in the case of the Continental Center, the deep girders allow the passage of ducts and pipes through their webs. The welded connections between girders and columns form a series of rigid portal frames, and the reinforced concrete floors act as horizontal diaphragms in providing further resistance to the turbulent winds of the narrow street canyons. Each floor is an open area without permanent obstructions, since the elevators and utilities of the Inland are located in a separate tower wholly outside the main enclosure. The slender tower is braced with full-length diagonals in all the outer bays and sheathed in stainless steel to distinguish this "servant" area from the glass-enclosed volume which it serves.

The greatest work of welded steel framing is the Chicago Civic Center, a skyscraper that for boldness and power stands in a class by itself (Fig. 69). The huge building was erected between 1963 and 1965 from designs prepared by three associated firms of architects and engineers—C. F. Murphy Associates; Skidmore, Owings and Merrill; and Loebl, Schlossman and Bennett. The building has only thirty-one floors other than those reserved for mechanical and electrical equipment, but it rises to a full height of 648 feet because of the extremely high ceiling height required by court and hearing rooms, auditoriums, and other public enclosures. The primary bearing members are sixteen columns of cruciform section, set for a bay span of 48 feet on the short elevations and 87 feet on the long—unheard-of dimensions that revolutionized the scale of building elements. The long span required floor girders in the form of the Warren truss 5 feet 4 inches deep; this depth is held uniform for the entire floor framing system, regardless of span, so that ducts and conduits can pass through the triangular openings between the web members of the truss. The welded connections between the trusses and columns make the primary frame a system of portal bracing, which is supplemented by K-truss bracing in the 29-foot bays of the elevator shafts. This apparently marks the first use of the K-truss in steel-frame buildings, although it was revived for the main trusses and lateral frames of bridges shortly after the turn of the century, nearly eighty years after Stephen Long invented it.

The external covering of the framing members in the Civic Center is a special kind of high-strength corrosion-resistant steel manufactured under the trade name of Cor-ten. It was developed in 1933 for use in railroad hopper cars, which are exposed to extreme conditions of abrasive and corrosive action. The steel was first applied to a structural frame in the administration building designed by Eero Saarinen for the John Deere Company at Moline, Illinois (1962–64). The unusual property of Cor-ten is that its oxide soon forms an impervious covering that prevents any further chemical action and that gradually darkens into a rich red-brown color. Because of these characteristics the exposed steel surfaces of the Deere building and the Civic Center have been left unpainted to provide a striking example of the union of economic and aesthetic virtues.

If a beam is welded at its ends to a pair of columns, or riveted to them throughout its depth, the rigid connection prevents it from rotating about either joint. Moreover, the column, being rigidly restrained by the beam and the adjacent leg, cannot be bent around its base by a horizontal force. If the beam is deflected, the bending force is transmitted to the columns; in other words, the stress in the beam is shared with the vertical members. Known as a rigid frame, such a system can be designed to offer a great advantage over the older arrangement: under a vertical load, the tensile and compressive stresses arising from the deflection of the beam are distributed uniformly among the three members rather than concentrated in the horizontal one. The origins of rigid framing as a special mode of structural design are not clear. Portal frames of any kind are in effect rigid frames, although for many years the complex pattern of bending in the rigid structure was not understood, and consequently they were not designed to realize the inherent efficiency of the form. An early (and possibly first) attempt to take advantage of the special behavior of the rigid frame ended in disaster. The roof of the Orpheum Theater in New York was intended to be supported on trusses riveted throughout the depth of the web to the columns; the expectation was that the stresses in the truss would be shared with the vertical supports. This was indeed the case, but the roof collapsed during construction in 1913 because the engineers had failed to understand the complex pattern of high stresses at the knee, where the horizontal and vertical members bend toward each other around the rigid connection. As far as the record shows, there was no further development in the United States for nearly two decades.

The first structure other than a bridge in which the rigid framing on a large scale was carefully designed according to its statical behavior has a special place in the history of American building. Cincinnati Union Terminal, built between 1929 and 1933, is the second-last of the great metropolitan stations in the United States, followed only by the Los Angeles terminal of 1939. It marked the end of a century of railroad dominance in the transportation industry, and yet it was the first to be designed without recourse to historical styles. The architects of this masterpiece of planning and structural ingenuity were Fellheimer and Wagner, and the chief engineer was Henry M. Waite. The main concourse of the terminal, elevated above the track level, extends back from the entrance lobby nearly at right angles to the axis of the track layout. The roof of the concourse is carried on fifteen rigid frames with a 78-foot span. The platform canopies, ordinarily most undistinguished examples of the builder's art, are carried on linear series of rigid frames which are arranged in tandem under the center lines of the roofs. These frames are joined in continuous units of three spans each, the individual span having a length of 80 feet. The steelwork is exposed, so that one can see in the downward-tapering legs how the bending moment decreases from a maximum at the knee to a theoretical zero at the foot.

The most elegant work of rigid-frame construction and the largest are both in Chicago. Crown Hall at the Illinois Institute of Technology is a leading example of the architecture of Ludwig Mies van der Rohe and his engineering associate, Frank J. Kornacker. Built between 1955 and 1956, Crown represents the ultimate refinement of form in a geometric rather than a structural sense. The building is a single enclosure of glass, 120 × 220 feet in plan, whose primary supporting elements are four welded rigid frames of strictly rectangular form in which the horizontal member stands clear above the roof to its full depth of six feet. The legs are H-columns with a uniform cross section throughout their height. These shapes were dictated by the aim of visual harmony and purity rather than the most efficient distribution of metal, which would have required that the horizontal member decrease in depth from the knees to the center and that the legs taper downward to a minimum section at the base. This latter profile characterizes the truss frames that support the roof of the Metropolitan Fair and Exposition Building in Chicago, another enclosure with an open interior but in this case one of record size (Fig. 70). Known locally as McCormick Place, the exposition building was constructed on the lakefront between 1958 and 1960, with Shaw, Metz and Dolio in charge of the architectural design, and Ralph H. Burke, Inc., responsible for the engineering work. This gigantic room measured 340 × 1,050 feet in over-all area; yet its reinforced concrete roof slab was carried on only eighteen frames, each spanning 210 feet clear. The longitudinal trusses connecting the frames, together with the bracing in the horizontal plane, made the roof structure a space frame, that is, a three-dimensional truss that acts as a unit (see pp. 198–200). McCormick Place lasted less than seven years: on January 16, 1967, a fire fed by combustible contents caused the steel frames to buckle, with a resultant collapse of roof and walls.

Associated with the rigid frame is the truss in which rigidity is secured through rigid connections rather than the triangular arrangement of members. Known as a Vierendeel truss after its inventor, this type has a history in bridge construction going back to the late nineteenth century (see pp. 224–25), but it was not used in building frames until the 1930's. The rectangular openings of the rigid-frame truss make it particularly suitable for the support of internal loads over long spans, since the openings in the web correspond in shape to the rectangular outlines of corridors and windows. Vierendeel trusses were confined to conventional building frames until about 1960, when engineers first recognized the possibility of constructing external walls as bearing-truss structures. The idea was given its initial embodiment in steel by Skidmore, Owings and Merrill when they designed the Beinecke Library at Yale University, built between 1962 and 1963 (Fig. 71). The main part of the library complex is a five-story open enclosure raised above grade level by pyramidal concrete piers located at its corners. In order to dispense with interior columns and to carry the 131-foot length of the block in a clear span between the piers, the walls of the enclosing structure had to be built as steel trusses. The architects accordingly designed each wall as a rigid-frame truss made up of rows of squares bounded by the arms of equilateral crosses welded together into a working unit. The areas defined by the arms of the crosses are covered with a curtain of marble slabs, and the steel truss is sheathed in granite. The external appearance of this sumptuous work thus suggests that it is a masonry prism, which is not only at odds with its underlying steel construction but would be in itself a structural absurdity.

The most important innovation in large-scale steel framing from the economic standpoint is the structural system known as the braced tubular cantilever. Developed primarily by the engineer Fazlur Khan in the Chicago office of Skidmore, Owings and Merrill, the form grew out of the problem of wind-bracing in extremely high buildings with light curtain walls of glass and wide bay spans. Khan reasoned that if a building could be constructed as a tubular framework, like the iron or steel bents of railway viaducts, it would behave like a vertical cantilever in which the bending and shearing forces of the wind would be sustained entirely by the rigid truss-like frame of the walls. A variety of such tubular cantilevers began to appear around 1965. The most spectacular is the 100-story John Hancock Building in Chicago, erected in 1965–70 from the designs of Skidmore, Owings and Merrill. The walls of the Hancock are X-braced trusses, much like the Howe truss of 1841. For the twin 110-story towers of the World Trade Center in New York (1968–73), the engineers Worthington, Skilling, Helle and Jackson, in association with the architects Minoru Yamasaki and Emery Roth, chose a Vierendeel truss with web members so densely arranged as to give the appearance at a distance of weightless screens. A structurally similar form is the massive Vierendeel trusses that constitute the walls of the Sears, Roebuck Building in Chicago (1970–75), another work of Skidmore, Owings and Merrill. The Sears comprises nine rigidly interconnected tubes which terminate in pairs at different levels, the two at the center rising to the record height of 1,450 feet. Since these buildings are cantilevers, under the deflection caused by wind loads, the windward side acts in tension and the leeward in compression, and the sides parallel to the wind direction take the shearing stress. The tubular cantilever has made it possible to reduce the weight of structural steel per square foot of floor area in a building by as much as 40 percent.

Curious forms of rigid-frame trusses appear in the walls of the International Business Machines Building in Pittsburgh—welded steel lattice trusses that bear a surprising resemblance to the original ancestor invented by Ithiel Town in 1820 (Fig. 72). A similar form in iron was developed by Jules Saulnier for the Menier Chocolate Works at Noisiel, France, as long ago as 1871. The architects of the Pittsburgh building were Curtis and Davis, and the engineers again Worthington, Skilling, Helle and Jackson. Since this is another structure in which interior columns are confined to the utility core, the engineers went to unusual lengths to provide rigidity against wind loads because of the flexibility of the thin-webbed lattice truss, a defect which plagued the timber bridges erected on the Town patent. Wind-bracing is divided among the welded truss walls together functioning as a rigid hollow box, by the reinforced concrete floors acting as horizontal diaphragms, and by horizontal trusses set in the floor planes to back the wall trusses. The last feature also has a long ancestry: an early example is in the Winthrop Building in Boston, constructed between 1893 and 1894. The serious limitations of the lattice truss and the long hiatuses in its use suggest that the IBM Building is something of a tour de force whose designers aimed at novelty of appearance as much as structural innovation for practical ends.

The great diversity of structural needs which the truss can satisfy suggested the possibility of uniting trusses lying in a parallel series or in two or more planes so that the resulting assemblage would act as a unit in supporting loads spread over a wide area. A three-dimensional arrangement of trusses is known as a space frame, and like the rigid frame, it existed in embryo for a long time before the engineers became aware of its potentialities. A truss bridge is in effect a prismatic space frame: the two parallel trusses of any one span do not act independently; they are rigidly tied together by top and bottom frames which are themselves trusses. The braced trestlework of timber, once common in American railroad bridges of the West, is a true space frame in which posts, longitudinal and transverse ties, and braces together form a three-dimensional truss. But again, as in the case of the rigid frame, engineers must create out of these strictly pragmatic responses to special conditions a structure consciously designed to exploit the full possibility of the technique. If loads are divided among members lying in three planes, the stresses in any one member can be drastically reduced; as a consequence, the number of members may be greatly increased but the individual pieces will be so small that their total weight can be kept relatively low and they can be assembled by hand-and-tool techniques.

The first space frame consciously designed for maximum rigidity and minimum weight was built in 1907 by Alexander Graham Bell at his summer estate at Braddock, Nova Scotia. The frame, which served only to support an observation platform, was a pyramidal tower of three legs, each of which was in itself a space frame in the shape of an elongated triangular prism. It was a perfectly rigid structure built up of short steel pieces bolted together. Although the whole assembly was remarkably strong for its light weight, there is no record that Bell's invention was put to further use or that it had any influence on the building arts. The space frame did not appear again in any American structure until the construction of George Washington Bridge between 1927 and 1931 (see pp. 234–36). The towers of this celebrated suspension bridge are two enormous rigid frames with an over-all height of 635 feet, constituted of three-dimensional trusses. Two-dimensional space frames—those in which the girders and beams of floor and roof frames are replaced by interconnected trusses—began to appear in the new single-floor assembly-line factories of the thirties. The potentiality of the form was advanced by two inventions that came around the time of World War II. In 1939 Charles W. Atwood of Detroit obtained a patent on a frame composed of a large number of small steel angles bolted together into a three-dimensional truss with the long axes in horizontal planes. Known as the Unistrut system, Atwood's invention was particularly suitable for the support of roofs with only a few widely spaced peripheral columns. A more radical innovation came in 1945, when the German engineer and theorist Konrad Wachsmann, in association with Paul Weidlinger of New York, received an American patent for a space frame of short steel or aluminum tubes joined by split-ring fasteners of a kind developed by Wachsmann. Called a Mobilar truss by the inventors, the frame is unlike any other truss arrangement in that a very large number of light pieces are disposed in a multiplicity of planes. Wachsmann's split-ring fasteners are capable of joining as many as twenty pieces in a single connection. He and Weidlinger proposed the Mobilar truss as a roof frame for an aircraft hangar, the frame being cantilevered outward 192 feet from a central support which is itself a space frame. Wachsmann became the most radical exponent of the industrialization of building techniques, and his inventions in space framing represent a long step toward the automated mass production of structural components.

The innovations in steel construction that we have so far considered are all rectilinear systems such as column-and-girder skeletons, trusses, and their derivatives, but the century has been equally prolific in the development of arched and other curvilinear forms appropriate to the framing of vaults and domes. The older forms of the hinged arch common to balloon train sheds, for example, found a new role in the construction of gymnasiums and drill halls. A particularly advanced structure in this specialized category is the Armory of the Twenty-Second Regiment in New York (1910–11). The vaulted roof of the drill hall is carried on twelve three-hinged arched trusses with a 200-foot clear span. The unusual feature of the frame, one without precedent in the construction of train sheds, is that the bottom hinges of the arches rest on girders cantilevered from the tops of the steel wall columns to provide maximum wall-to-wall clearance at the floor level. This method of support precluded the use of ties, with the result that the horizontal thrust of the arch had to be resisted by the increased depth of the column. In other respects the Armory followed balloon-shed construction, as in the use, for example, of longitudinal trusses that functioned as wind-bracing and roof purlins.

The hinged arch reached its greatest linear dimensions in connection with a once spectacular and now almost forgotten phase of transportation technology. The dirigible aircraft, or zeppelin, required hangars whose over-all size exceeded that of the largest train sheds. The first was built at the Naval Station at Lakehurst, New Jersey, in 1920, as a parabolic vault that mesaured 262 × 803 feet in plan and stood 172 feet high at the crown. The structural frame followed the principle embodied in the New York armory: three-hinged arched trusses with a rise of 108 feet rested on framed bents that lifted the bottom hinges 64 feet above the floor of the hangar. These bents were space frames in the shape of high slender pyramids. The crash of the American zeppelin *Macon* in 1935 and the burning of the German *Hindenburg* in 1937 ended the active life of the rigid airship less than thirty years after it had begun. But the non-rigid craft, commonly known as a blimp, proved useful for coastal patrol during World War II, and the United States Navy accordingly built a number of hangars on the Atlantic and Pacific coasts to house these ungainly ships. The largest hangar, designed by A. Amirikian and built at a naval air field in New Jersey during the early years of the war, had an over-all span of 328 feet and an interior height of 184 feet, both records at the time. The main supporting members of this enormous vault were two-hinged arched trusses tied at the hinges by a steel rod that was set a little below the operating floor.

The once valuable technique of hinged-arch framing lost popularity after World War II. The two-hinged form remains common for arch bridges of steel, but it has been largely superseded for domes and vaults by more efficient and economical structures such as concrete shells, welded steel-rib framing, and welded plate. The inherent disadvantages of the older technique came to be all the more pronounced in the face of these new structural forms: the erection of the hinged arch requires expensive centering and falsework, whereas welded ribs can be built out as cantilevers, and the costly formwork required for the concrete shell is offset by the great economy of the thin section. And by a curious irony of progress, the chief virtue of the three-hinged arch, its determinacy, has declined in importance because the new methods of stress analysis in which mathematical operations are performed by electronic computers have made the calculation of stresses in indeterminate structures a relatively easy matter.

The earlier forms of iron ribs or trusses developed to support the domes of monumental buildings began to disappear as the architecture that required them lost favor before newer fashions. The last building in which the dome was consciously adopted for monumental effect is Cincinnati Union Terminal (1929–33). The great semidome that faces the long terminal plaza is the first to embody a structural system that has no exact precedent in the long history of the form. Instead of being supported by radial members, the roof is carried by eight semicircular arched trusses of diminishing radius that lie in a series of vertical planes, one behind the other from front to rear. The engineers chose this massive and inelegant system of riveted steel trusses for several reasons: ribbed framing in a semidome would have been inadequately supported at the front wall, where the dome is cut in half; the ring foundation and the necessary buttresses or ties would have interfered with the vehicular passages extending under and around the dome; and a number of the ribs could not have been carried to the foundation because of the need for a wide opening between the rotunda and the train concourse. It was a bridge-builder's solution to a peculiar problem, ingenious and perfectly sound, but outside the main line of structural evolution. The fundamental idea survived, however, to re-appear in the timber trusses of outdoor band shells, which are geometrically similar to half-domes.

Although the dome disappeared as an element in a formal composition, it was to be reborn in a variety of functional designs intended simply to shelter the maximum area at minimum cost. One of the many consequences of the post-war flight of urban dwellers to the suburbs has been a strenuous competition among cities to attract conventions, exhibitions, and professional sports. A minor element in the total economic pattern of the nation was the expansion of the major baseball leagues from eight to ten teams, but it was to have a powerful effect on the building arts as cities vied with each other to guarantee profitable seasons to the owners. The special problem of combining auditorium, sports stadium, and exhibition hall in one enclosure stimulated the skill and ingenuity of engineers and builders. The flattened wide-span dome is the only structure that can provide complete shelter without intermediate supports for areas measurable in acres. The largest domed shelter of this kind so far built in the United States is the Harris County Stadium or "Astrodome," at Houston, Texas, which was constructed between 1962 and 1964 to provide air-conditioned seating for sixty-six thousand people (Fig. 73). Both the stadium proper and the associated playing field lie under a domed roof with a clear interior span of 642 feet that covers nearly seven and a half acres. A number of talents combined to produce this mammoth, an admirable work of structural science if not of architecture: Praeger, Kavanaugh and Waterbury were the structural engineers, and Lloyd and Morgan, associated with Wilson, Morris, Crain and Anderson, were the architects. The roof, half plastic skylights and half wood-fiber concrete boards, imposes a moderate unit dead load, but it was framed to sustain a wind load produced by a steady hurricane of 135 miles per hour and by intermittent gusts of 165 miles per hour. Structures of this size are comparable to wholly novel forms in the unpredictability of their behavior, and the engineers, accordingly, do not rely exclusively on mathematics and theory. A model of the Houston dome, built to a scale of one-eighth inch to the foot, was carefully tested in a wind tunnel before construction began.

The supporting frame of the dome is a complex arrangement of steel trusses in which the basic radial system is supplemented by circular and skew patterns. In this dense arrangement of members, stresses arising from uniformly distributed loads are quickly distributed through the entire framework and produce mainly axial forces with negligible bending. The primary bearing members are twelve radial trusses set at 30-degree intervals and spanning from a compression ring at the crown to a tension ring in the form of a circular truss at the periphery of the dome, where the horizontal thrust is translated into tension in the bottom ring. The secondary bearing system is made up of two main components: first, six circular trusses which join the trussed ribs and lie between the crown and the periphery of the dome; second, a dense array of trusses set in a herringbone pattern superimposed on the radial system of the ribs. All trusses are held to a uniform depth of 5 feet. The entire weight of the roof decking and the steel framing—3,750 tons—is sustained by the seventy-two columns on which the tension ring rests. The fundamental principle underlying this intricate system is that of distributing loads uniformly over a large number of relatively small members, so that each plays a small but vital role in the total action. It is the same principle that is embodied in the space frame; we will see it again in geodesic domes and in other structures appropriate to concrete and wood. It is an essential idea in contemporary building, and like the shells and prestressing in concrete construction, it is one of the primary features that distinguish the structural arts of the mid-twentieth century from the most advanced systems of the nineteenth.

Outside the ruling mode of framed construction there are techniques that exploit the physical properties of highly specialized shapes. Metal shells made up of riveted or welded plates, for example, form a structural category that is unrelated to any other kind of metal construction because they are continuous forms rather than forms made up of discrete elements. Shells are structures rigid in themselves, but they have certain characteristics in common with membranes and stressed skins, which can act only in tension because of their extreme flexibility. The basic principle of metal-shell construction affords another example of the scientific exactitude of contemporary building. Certain functional requirements can be satisfied most efficiently if the distinguishing property of a structural element is made to work to the maximum degree possible within its own limits. The shell, for example, is rigid enough to support itself and its internal load but is extremely susceptible to buckling or caving in. If the forces acting on it can be translated almost entirely into tension, the shell is the ideal answer to a certain kind of structural need. The problem is to find those conditions which enable the engineer to take advantage of the special property without running the risk of involving the weaknesses.

Metal shells have their roots in commonplace forms that began to emerge in the early period of iron technology. Boilers, water tanks, and gas holders are examples of shell construction, and their ancestry may be traced back to the origins of wooden casks. Steamships with iron hulls are shells, but these are reinforced with ribs to sustain the variety of internal and external forces acting on them. In all these cases there is a common factor: the primary load on the shell is a fluid pressure that either is distributed over the entire area or acts along any one section of the entire shell and that increases continuously in a certain direction from zero to a maximum. The most economical form for a fluid container is the sphere, since it has a minimum area for a given volume. For this reason the tank that holds a local water supply is now frequently built in spherical form, whereas it was once universally cylindrical. Because the pressure of a fluid increases with increasing depth, however, the sphere does not provide the mose effective distribution of material. The perfectly organic form for uniformity of load distribution and hence uniformity of stress is a compound spheroid like a soap bubble, somewhat elongated at the top and flattened at the bottom. The translation of this organic form into metal was achieved in 1928 by George T. Horton, who designed the first spheroidal steel tank, which was used for the storage of oil. These squat pumpkin-like shapes are now common on the tank farms of the petroleum industry. Rigid metal shells have a limited application in buildings and have so far been confined chiefly to nuclear power plants. An early example is the reactor housing of the Commonwealth Edison Company's nuclear generating station at Dresden, Illinois (1958–60). The housing is a metal-plate sphere, internally self-supporting, with a diameter of 190 feet.

The idea of a stretched membrane (usually called a stressed skin) as the protective covering for a building was first proposed by Richard Buckminister Fuller in the prefabricated house that he patented in 1946. The covering was to be a thin aluminum skin streteched over a light framework of ribs, combining curtain wall and roof in a single continuous enclosure. The house was among the inventions by means of which Fuller hoped to bring about the industrialization of the building process. Every part of the structure and all fixed interior utilities were to be prefabricated by mass-production methods and packaged ready for assembly on the site. By using aluminum and other light-weight materials, Fuller held the total weight of the structure, the wall and roof covering, and the interior partitions to eight thousand pounds. But this highly rationalized adaptation of building to industrial processes was too radically at odds with conventional standards of form and construction for Fuller to find a sponsor, and the house never reached the market.

The greatest work of stressed-skin construction so far undertaken is the Gateway Arch in St. Louis, erected between 1962 and 1966 as America's boldest and highest monument. The winning architect in the competition for this design was Eero Saarinen, and the structural engineers were Severud, Perrone, Fischer, Conlin, and Bandel of New York. The arch was conceived as the Gateway to the West, the central feature and primary symbol of the Jefferson National Expansion Memorial, which was established by executive order in 1935 but remained unrealized until it was included in the National Park Service's conservation and development program known as Mission 66 (1956–66). In addition to the arch, the Memorial includes a fifty-acre landscaped area that was once the granite-paved levee at the Mississippi wharfs, an underground museum and visitors' center, and the old Court House and Cathedral, now fully restored (see pp. 69–70, 87–90).

The Gateway Arch is an isolated structure springing from grade level, rising to an over-all height of 630 feet, and spanning an equal distance between the outside surfaces at the base. Saarinen designed it as an inverted catenary in profile and thus gave it the most efficient form since the axis corresponds exactly to the pressure line under a load uniformly distributed along the curve; this is the case here because the vertical load is simply the dead weight of the structure. The cross-sectional shape is a double-walled equilateral triangle in which the hollow core provides a passageway for the elevator train that carries passengers to the observation gallery at the crown. The exterior walls are a stressed stainless-steel skin one-fourth inch thick; the inner walls are built up of three-eighths-inch carbon-steel plate. Rigidity is provided by a vast number of little steel cells between the two walls and by a reinforced concrete infilling extending up to the 300-foot level. Above this point the walls are stiffened wholly by the cellular construction, a technique first used in the United States for the towers of Golden Gate Bridge (see pp. 237–38). All steel members of the St. Louis arch are joined by welding. The concrete and cellular backing was designed to withstand a horizontal pressure induced by winds of 150 miles per hour, at which velocity the deflection at the crown from the vertical plane is calculated to be 18 inches. The abutments of the arch are solid concrete blocks set on a sloping axis and extending 60 feet below grade, well into the limestone bedrock of the St. Louis area.

The erection of the Gateway Arch was a painfully tedious process, several times delayed by difficulties inherent in the process itself. The outer and inner walls were prefabricated in relatively small sections and bolted together at the factories of the fabricating and erecting contractor. These sections were shipped by rail to St. Louis and welded on the site into equilateral-triangular prisms. The finished fifty-ton sections were placed up to a height of 72 feet by cranes located on the ground; above this level they were placed by climbing cranes that crawled like insects up the outer surface of the arch. This process continued up to the crown or keystone section, where the observation gallery is located. The final section was placed by forcing apart the two legs by an eighty-ton scissors jack and setting the crown section between them. This process, adapted from arch bridge construction, places the outer surface of the arch under compression, which offsets the tensile stresses arising from bending induced by wind loads.

Another technique of twentieth century building that grows out of the primary physical property of a special metal shape is suspended construction. Like many other innovations in the structural arts, its roots lie in the nineteenth century. The chief source is the wire-cable suspension bridge, which had reached a high stage of development by 1860. Since the wire cable has an extremely high tensile strength but is unable to sustain any compressive load, its very flexibility gives it certain advantages over rigid members, provided that it is constrained to act only in tension. The first to propose the application of suspended construction to buildings was James Bogardus. He submitted a project for the Crystal Palace of the New York Exhibition of 1853 in which the roof of a circular cast-iron building 700 feet in diameter was to be suspended from radiating chains anchored to a central tower. The second proposal for a suspended structure came forty years later when the architect E. S. Jenison offered a plan for an enormous multifloored circular building to house exhibits at the Columbian Exposition of 1893, the entire enclosure to be suspended from a central mast 1,100 feet high by means of radiating cables. On a more modest scale, Buckminster Fuller in 1927 obtained a patent for what he called his Dymaxion house, the roof of which was suspended from six wires fixed to a central mast. This was the first of Fuller's attempts to adapt construction to industrial processes, but like the 1946 invention, it failed to reach a market.

The Bogardus and Fuller projects placed the support for the suspension cables at the center of the structure, which made it unnecessary to include separate anchors for the cables or peripheral supports for the outer edge of the roof. When the idea finally reached the stage of realization, however, the designers reversed this relationship and located the masts on the periphery of the enclosure. The translation of the theory into actuality came with the Travel and Transport Building at Chicago's Century of Progress Exposition (1933–34). One of the most important of the many structural innovations at the fair, the building was the work of the architects Edward Bennett, Hubert Burnham, and John Holabird, and the engineers B. M. Thorud and Leon S. Moisseiff. Although it was a sensational piece of "modernistic" fair architecture, its designers made no effort to hide the elaborate suspension structure (Fig. 74). The domed roof of the circular building was suspended from cables fixed to a circumferential ring of twelve steel towers 150 feet high. The tension in the supporting cables was balanced by backstays anchored to a ring of twelve concrete blocks that were imbedded in the ground outside the building wall. The towers were wind-braced in the vertical plane by means of portal trusses that tied them together in groups of four. Although a clumsy and redundant system by later standards, the Transport Building represented a valid embodiment of the principle of restricting the thin cable to direct tension and the column to compression. The chief purpose in adopting this design was to eliminate the heavy steel ribs of conventional dome framing, with the attendant high cost of falsework. The engineers of the building, however, failed to exploit fully the potentialities of suspended construction because the system required them to include separate backstays and anchors and to retain at least a light framework of ribs and purlins under the dome.

In the mature form of suspended construction the roof deck is supported entirely by the cables; this requires that the usual form of the dome be inverted, since the flexible cables naturally sag downward and cannot be made to bow upward. The emancipation from conventional forms came with the construction of the State Fair and Exposition Building at Raleigh, North Carolina, between 1954 and 1955. The architect of this important pioneer work was William H. Dietrick and the engineers were Severud, Elstad and Krueger; the idea of the suspended roof, however, was first proposed by the Polish architect Matthew Nowicki. The roof decking of the Raleigh building rests on a two-layered grid of cables strung at right angles within a ring beam composed of two sloping parabolic arches opening toward each other on a common axis and intersecting in hinged joints at the low point of the roof. The shape of the arches and the location of the cables give the roof the form of a hyperbolic paraboloid. Although this complex surface is curved in each of the three perpendicular sections, it is nevertheless a ruled surface, that is, one that is generated by straight lines, which means that the surface can be formed by a network of straight, tightly drawn cables. Since these flexible elements are subject to pure tension, the ring beam, being drawn toward the center, is theoretically limited to compression, but some bending arises from wind loads and buckling tendencies. The resulting tension and shear are taken by the reinforcing in the concrete arches, which are partly supported by steel columns in compression.

Thus each element works to maximum efficiency, being restricted exactly to the kind of load for which it is best suited.

The cables of a suspended roof may be arranged in a radial system, which offers certain economic advantages: the cable lengths are uniform, the tension is in theory distributed equally among them, and the compression in the ring beam is uniform throughout its length. In a radial pattern the cables are fixed at their inner ends to a small central ring that must necessarily be under tension. But the very simplicity of the radial scheme brings to an acute stage the inherent difficulty in suspended construction. Under wind loads the thin flexible cables vibrate rapidly in the vertical plane, and this flutter, as it is called, may reach dangerous proportions where there is only a single layer of cables. One solution, suitable for very large installations, was developed by the engineer Lev Zetlin in 1958, when he introduced a double set of cables arranged like the spokes of a bicycle wheel and added rigid stiffening bars between pairs of cables in the vertical planes. For smaller spans the single system proves satisfactory if the cables are intimately associated with a roof that forms a continuous web between the tension and compression rings. A radial pattern of this kind characterizes the roof structure of the circular restaurant located between the two domestic terminal buildings at O'Hare Field in Chicago, designed by the architectural and engineering firm of C. F. Murphy Associates and opened to the public in 1962. Another form of suspended construction, with heavy precast concrete decking, is the one-way cable system supporting the roof of the terminal building at Dulles International Airport, Washington, D.C., built between 1960 and 1962 after the plans of the architects Eero Saarinen Associates and the engineers Ammann and Whitney.

One of the few works of suspended construction to follow Buckminster Fuller's original form of central mast and downward-radiating cables is the East Memorial School at Greeley, Colorado, built between 1963 and 1964 from the plans of the architects Shaver and Company and the engineers Bob D. Campbell and Company. This small elementary school consists of three hexagonal classrooms bounded by a single hexagonal enclosure open at the top. The roof frame of each of the rooms is suspended from twelve galvanized steel rods radiating downward from a central mast in the form of a steel box column. Six of these rods are joined to the corners of the hexagon, and six drop to interior points on the radial beams. The light construction of the roof requires that it be held down as well as supported from above. The anchors are three-inch diameter pipes located at the corners of the hexagon, their lower ends imbedded in concrete blocks.

The most widely publicized structural invention of the century is one that has no exact antecedent nor parallel in the building arts. The geodesic dome, patented in 1947 by Buckminster Fuller, is the equivalent in arched forms of the light space frame among rectilinear systems. Fuller's primary intention was to substitute a great number of light, mass-produced, hand-assembled elements for the heavy ribs and purlins of the conventional steel-framed dome. In this respect he succeeded admirably: for relatively large enclosures, he was able to reduce the weight to as little as 1/300th of that of the traditional forms. The finished geodesic dome, which is approximately spherical in shape, is a thin cover of aluminum or plastic pinned to a framework of steel or aluminum tubing that is arranged in a pattern of little triangles or hexagons, more often the latter. The connections are either bolted or made by a special kind of joint in which the tubes are inserted into small holes drilled in a circular disc. Since the triangle is a rigid figure, and contiguous hexagons rigidly support each other, the whole domed structure is a rigid framework theoretically limited to compressive stresses. Because the separate elements are straight lengths, the actual shape of the dome is a polyhedron with a great number of faces—nearly a thousand of them in large industrial installations. Further, because the maximum number of faces for a regular polyhedron is twenty, it is impossible to construct most geodesics of perfectly uniform pieces. But this is a minor defect, more than offset by the great virtues of the technique. Geodesic domes of near-record size have been built from standard tubing and the most ordinary kinds of hardware, all the elements mass-produced at minimum cost and quickly assembled by common hand-and-tool skills.

For the first decade of its history the geodesic dome was limited to small structures, most of them built for experimental purposes at schools of architecture and engineering. This situation was drastically changed in 1958, when the engineers Battey and Childs designed an immense dome to house the car-repair facilities of the Union Tank Car Company at Baton Rouge, Louisiana. Although the dome is 384 feet in diameter, with a clear interior height of 120 feet, the steel tubing and connections weigh only 567 tons. The protective covering is a continuous aluminum skin pinned to the underside of the framework in a pattern of flattened pyramids, each connected at its apex to the center point of the hexagonal cell and held in place by six short rods that radiate from the center to the corners of the hexagon. Because the little pyramids of aluminum tend to pull these rods downward, they are subject to tension, whereas the structural tubes that form the hexagonal cells are normally in compression.

The Union Tank Car Company revolutionized the method of erecting domes when they built their second geodesic structure between 1959 and 1960, to house the company's repair shop at Wood River, Illinois. The dome frame of one thousand hexagonal cells was assembled from the top down on the surface of a gigantic balloon inflated to a maximum pressure of only 0.1 pound per square inch above atmospheric pressure. As each course of hexagons was completed, the balloon was expanded sufficiently to make room for the course immediately below it. The technique spelled the end of falsework for geodesic structures and thus considerably enhanced their already superior economy. Their wide range of uses was further extended by the construction of the double dome, one inside the other, built between 1961 and 1962 to house the botanical conservatory at Shaw Gardens in St. Louis, Missouri. This is also the first large installation in which the protective sheathing is a transparent plastic skin that can admit solar radiation.

As in the case of the Unistrut and Mobilar frames, Fuller's invention extends the area of industrialized building, and the inventor confidently looks forward to the day when all mankind will be securely and comfortably housed in separate domed enclosures mass-produced at $7,000 a dwelling unit (1964 price level). There is little question that this is technically feasible; but as a social philosopher, Fuller has exhibited an unshakable naïvete that blinds him to the social and economic problems that stand in the way of this utopian realization, even if we grant that such a mechanized solution is desirable. Another fact that we must bear in mind in the face of the philosophical enthusiasms the geodesic dome has generated is the inability of the light frame to carry superimposed loads. Since it is limited to the support of a protective cover, it can play no role in the construction of the great majority of urban buildings and consequently cannot be said to have brought about the architectural revolution that the public and professional press has so often proclaimed.

Chapter Seventeen | Steel Bridges

Although steel-bridge construction in the United States during the twentieth century includes many imaginative and spectacular works, it has not been characterized by the innovative spirit revealed by other areas of building. And yet, ironically enough, many of the inventions incorporated in buildings were derived from the structural forms of bridges. The chief tendency since 1900 in the design of truss bridges has been the reduction of the web system to posts and single diagonals, as in the Pratt and Warren trusses, but this has not always been possible in the largest railroad bridges, where subdivided panels are frequently unavoidable because of the necessity to distribute the heavy concentrated loads as uniformly as possible over the bottom chord and to minimize the buckling tendency in web members. The growing refinement of form has come about through increasing scientific and mathematical rigor in design and through the load and wind testing of models; yet such progress would have been impossible without a parallel improvement in the quality of steel, accompanied by a steady reduction in its price. Bridge construction throughout the first quarter of the century continued to be dominated by the railroad as the nation's rail network and its traffic steadily expanded to a high point around 1920. With increasing traffic went increasing locomotive weight, which reached a maximum at the very end of the development of steam power, before the diesel-electric engine supplanted it. After 1925, however, the rapid growth of automotive transport soon reached the point where the very longest spans and the overwhelming majority of all bridges were built for the highway. This revolution in transportation shifted the focus throughout a wide range of the building arts, from rail to roadway bridge, from railroad terminal to airport, and from freight yard to parking lot and garage. And with these have come a host of urban problems not even foreseen when the railroad was the most voracious devourer of land.

The simple truss reached its greatest length in the bridge of the Burlington Railroad at Metropolis, Illinois, and it was again the Ohio River that established the need (Fig. 75). Built between 1914 and 1917 under the design and supervision of Ralph Modjeski and C. H. Cartlidge, this big but sober and conventional bridge united a long branch line of the Burlington with the Nashville, Chattanooga and

St. Louis Railway at Paducah, Kentucky, and thus established a new freight route that joined the midwestern prairies with the southeastern states and the Gulf Coast. The bridge is also a vital link in the Illinois Central Railroad freight line known as Edgewood Cut-Off. The over-all length of the river crossing at Metropolis is nearly 3,500 feet, divided into seven spans, of which the channel span has the record length of 723 feet. The supporting structure is a pair of through Pratt trusses with subdivided panels and a polygonal top chord, a form derived from the Ohio River bridge of the C. and O. Railway at Cincinnati (1886–88; see p. 143). Pier construction at Metropolis was a relatively simple matter, free of the problems usually encountered in foundation work under broad navigable waterways. The concrete piers rest on a thick bed of compact gravel near enough to the surface of the river bed so that it could be reached by open excavation within cofferdams.

The Warren truss offers the possibility of greater elegance than the Pratt because the diagonals form a continuous pattern, although the subdivisions common to rail spans tend to interrupt this continuity. The more refined appearance is especially true of the deck truss, which is often built with parallel chords and thus avoids the humped profile of the polygonal chord in the through truss. Among railroad bridges in this category, perhaps the handsomest is the five-span structure built between 1937 and 1939 to carry the main line of the Chicago, Rock Island and Pacific Railroad over the Cimarron River near Liberal, Kansas. Designed and erected by the company's Bridge Department, the span is the central feature of a cut-off built to avoid the grades and curves of the old descent into the river valley. The long horizontal lines of the top and bottom chords of the deck trusses are enhanced by the sharp-edged piers, each of which is divided into a pair of concrete shafts with a square cross section.

Although the simple Pratt or Warren truss with subdivided panels remained the standard for long-span bridges, the continuous truss was beginning to come into vogue around the turn of the century. For most engineers the peculiar problems associated with this form more than offset its inherent advantages. After 1900 its indeterminacy was no longer a serious matter, since a number of valid methods of stress analysis had been developed prior to that date, but there were other difficulties. Pier settlement may produce secondary stresses (those arising from factors other than loads), stresses unwanted in themselves and unpredictable in their effects. Other secondary stresses, such as those induced by thermal expansion, common to all structures of metal or concrete, are cumulative in continuous trusses and thus may reach dangerous proportions. Finally, the rhythmically changing bending moments reach a maximum level directly over the supports. But continuity possesses attractive virtues that outweigh the defects if the engineer can find a way of circumventing them. Since the continuous truss is not divided into separate units, as is the case with a multispan bridge of simple trusses, the rigidity of the whole structure is very much greater than that of the older form; moreover, since the whole structure works as a unit, the load on any one panel will be transmitted to all the others and reduced in its effect by being shared among them.

Up to the time of World War I continuous structures had been confined to girder bridges and to short-span trussed forms, such as those built for metropolitan elevated lines. This limited role was suddenly expanded when Gustav Lindenthal proposed continuous trusses for the Ohio River bridge built by the C. and O. Railway at Sciotoville, Ohio, between 1914 and 1917. This daring and imaginative engineer was at the time winning international attention for his Hell Gate Bridge at New York (see pp. 229–31), and the simultaneous construction of the bold span at Sciotoville was to establish him once and for all in the front rank. The two parallel trusses that carry the double-track railroad line over the Ohio River remain the largest continuous trusses ever built.

There was good reason for this expenditure of talent and money in a little community a few miles east of Portsmouth, Ohio. The main line of the railroad lies on the south bank of the Ohio River as far as Covington, Kentucky, where it turns sharply to make the crossing into its former western terminal at Cincinnati. But the company's heavy and rapidly expanding coal traffic has always been destined in good part for the industrial centers of the Great Lakes, especially Detroit, Toledo, and Cleveland, which could be reached only by shipment over other roads on comparatively circuitous routes via the Cincinnati gateway. The Ohio River bridge (1888), however, and the tangled pattern of yards and terminals made Cincinnati more a bottleneck than a gateway. A direct route to Toledo and Detroit was the obvious answer. The C. and O. took the first step in this direction by gaining control of the Hocking Valley Railroad, whose main line extended from Columbus, Ohio, to Toledo. The program was fully realized by constructing a new line extending south from Columbus to the river and by building the bridge necessary to reach the original line on the south bank.

The entire length of the river crossing at Sciotoville is supported by a single pair of continuous trusses extending 1,550 feet in length (Fig. 76). Because the channel shifts from the Kentucky to the Ohio side during times of high water, the Army Engineers required that this length be exactly divided by a center pier into two clear spans of 775 feet each. The huge truss, a subdivided Warren with a maximum depth of 129 feet over the pier, is thus perfectly symmetrical, a characteristic that is intensified by the balanced deck-truss approaches. The danger of secondary stresses arising from pier settlement could be avoided because bedrock coincides very nearly with the surface of the river bed. Other unusual features combine to give the bridge a unique place among American railroad spans. The end bracing matches the mass of the main truss: a portal arch with a 6-foot depth at the crown is surmounted by a portal truss 23 feet deep. The main floor beams have the form of inverted U-shaped rigid frames whose vertical members are riveted to the posts of the main truss up to the lower chords of the lateral trusses that constitute the primary bracing elements in the top frame. The whole structure represents an extraordinary union of great weight and rigidity with a highly articulated framing system. Confronted by large bridges remote from the urban area, we lose our sense of scale, with the result that the Sciotoville span seen in side elevation suggests open articulation rather than weight; but at close range and with a train on it to suggest its true size, it conveys an unparalleled sense of mass and power.

The American classic among railroad bridges is the continuous-truss span built by the Texas and New Orleans Railroad between 1942 and 1944 to carry its main line over the Pecos River near Langtry, Texas (Fig. 77). The new structure replaced the precarious-looking cantilevers that had been erected a half-century earlier (see pp. 147–48). The continuous Warren truss, erected without subdivisions and hence reduced to its ultimate simplicity, carries the rail at a height of 321 feet above mean water level. The top chord of the deck truss is of necessity horizontal, while the profile of the lower chord corresponds to a certain extent to the changing bending moment over the length of the bridge, which reaches a maximum above the intermediate piers. The geometric purity of this form, the exact symmetry of the seven-span length, and the bare, fantastic beauty of the West Texas desert all combine to give the Pecos River bridge its powerful visual appeal.

The effective use of the cantilever bridge to support rail loads over a span of nearly 800 feet meant that it could easily be built to carry highway loads over much greater lengths. Enthusiasm was understandably tempered by the collapse during construction in 1907 of one of the cantilever arms of the first St. Lawrence River bridge at Quebec, a catastrophe that cost eighty-two lives. But construction at the same site was to begin again four years later and to end successfully in 1917 as the longest trussed span in the world. Meanwhile, the cantilever bridge was growing to new records in the United States. The Queensboro Bridge in New York, constructed between 1901 and 1908, is the third of the eight East River bridges and the first to be built in other than the suspension form. It is also the first major work of Gustav Lindenthal, who exhibited his characteristic boldness by stretching the larger of the two river spans to a length of 1,182 feet. Queensboro Bridge, being an extension of East Sixtieth Street on Manhattan, nearly bisects Blackwell's Island in the East River. Lindenthal took advantage of the land at mid-stream to divide the river crossing into two long spans above the water and one short length above the island, the latter acting as a common anchor for the two cantilevers on either side of it. An unusual feature of the long spans is that they are composed of pure cantilevers without the customary suspended structure between them. This is indicated by the uniform arrangement of the diagonals in the subdivided Pratt trusses and by the profile of the top chord, which descends steadily from its high point over the piers to a low point exactly at the mid-span, reflecting the uninterrupted decrease of the bending moment from tower to mid-span. Queensboro is a work of expert engineering, but its harsh angularity and awkward profile make it one of the less attractive of large urban bridges.

The next cantilever span in the same size range as the New York structure includes a number of innovations dictated by the special conditions of its site. Carquinez Strait Bridge at Crockett, California, designed by David B. Steinman and William H. Burr, is the first of the great complex of bridges that knit together the scattered cities in the area of San Francisco Bay. The need for bridges in regions of broad waterways and the far-reaching role they play in changing social and economic patterns are nowhere better demonstrated than in the Bay area. San Francisco was the most isolated of large cities not located on an island, for its narrow peninsula provided the only land access to the rest of the state. East, north, and northeast of it are broad tidal embayments and estuaries formed when the old valleys and canyons were drowned by the subsidence of the coastal lands. The line of the Southern Pacific to Los Angeles is the only track that enters the city itself; the other rail lines terminate at Oakland, where passengers and freight had to be ferried over the five miles of intervening water. East-west highway access to the city was cut off in the same way, as was the system uniting the north and south halves of coastal California, in the latter case by Suisun and San Pablo bays. Although the breadth and depth of the waterways and the swift tidal currents that flow in their narrower reaches were alone sufficient to discourage the most ambitious builder, the possibility of earthquakes added a seemingly insurmountable obstacle to the construction of bridges. Neither the engineering nor the financial resources were adequate for the job.

After 1920, however, the situation began to change with the rapid growth of automobile traffic, which not only made the need crucial but also offered the possibility of satisfying it. The decisive step came in 1923 with the establishment of the American Toll Bridge Company for the purpose of raising the necessary capital by borrowing against the expectation of steadily increasing tolls. The bridge was placed under construction in the same year and opened to traffic in

1927. The logical place to begin the assault on the waterways of the Bay area is Carquinez Strait, the relatively narrow body of water separating Suisun and San Pablo bays. The bridge that Steinman and Burr designed, although conventional in over-all form, contains a number of novel features incorporated to help the structure resist earthquake shocks. The main part of the crossing over water is divided into two long spans composed of the familiar cantilever-suspended combinations, the longer of the two, over the channel, extending 1,100 feet between bearings. The Pratt trusses are subdivided in an unusually elaborate way in order to distribute the impact of a seismic disturbance throughout a large number of members. The distribution of stress is insured by tying all the spans rigidly together into what is in effect a continuous truss. The unique elements in these protective measures are the hydraulic buffers at the expansion joints, which were introduced to damp down extreme shocks followed by oscillatory movements (an exaggerated counterpart of vibrations that may be induced by wind). Finally, in place of the usual concrete piers, the engineers substituted steel bents in order to take advantage of the elasticity of the metal rather than rely on the unyielding rigidity of the concrete. The center bent is a tower in the form of a space frame whose longitudinal depth is greater than the width of the deck and which acts as support, anchor, and tie for the contiguous cantilevers. The construction of the piers under the bents was plagued by troubles: the tidal currents tipped the first caisson 13 feet out of plumb, and the cofferdam for the big tower pier was damaged so badly by the wood-boring clam *Teredo* as to be rendered useless. The bridge has served well since its completion, although it has not been seriously tested by the special forces it was designed to withstand.

Although the Carquinez bridge made a valuable contribution to structural techniques, it did little to improve the appearance of the cantilever truss. The first bridges to reveal the potential grace of this ordinarily awkward form are the identical structures built to carry Pulaski Skyway over the Passaic and Hackensack rivers between Newark and Jersey City, New Jersey (Fig. 78). Constructed between 1930 and 1932, the Skyway elevated Lincoln Highway above the industrial tangle of the North Jersey meadows in order to provide a direct approach to the west end of Holland Tunnel (1927–31). Most of the three-mile length of the Skyway is carried on continuous deck trusses of short span, but at the rivers these give way to swinging anchor and cantilever trusses of unusually pleasing form for this type of bridge. The chief engineer of the project, Sigvald Johanneson, introduced a number of modifications into the commonplace cantilever that were in part dictated by necessity and in part made by deliberate choice from among possible alternatives. The most conspicuous features are the swinging anchor trusses that curve upward from the deck to the through position to provide an easy transition from the approaches to the main span. The abrupt changes in the top-chord profile of the river trusses were softened by dropping the lower chord at the piers as well as raising the upper chord to secure adequate truss depth at this point of maximum bending. Having decided on this feature, however, and being faced with the necessity of providing full clearance for nearly the entire width of the waterway, Johanneson had to use very short cantilevers and comparatively long suspended spans (75 feet for each cantilever against 400 feet for the long truss). This decision further improved the appearance of the bridges by stretching out the curve of the top chord in the suspended span.

The conditions that dictate the form of the largest cantilever spans seldom allow any latitude to the designer, who is closely bound by economic and structural requirements. This is obviously the case with the two tremendous structures that dominated truss-bridge construction in the prolific decade of the thirties, the East Bay crossing of San Francisco-Oakland Bay Bridge and the Huey P. Long Bridge at New Orleans. They represent straightforward engineering, stunning for their size and their triumphs over natural obstacles, but squarely in the main tradition. The double-deck Bay bridge (1933–37) embraces two different structures physically separated by Yerba Buena Island. The West Bay crossing is a double suspension bridge (see pp. 236–37), whereas the eastern portion is an inharmonious combination of through and deck trusses, the various parts having an over-all length of 8¼ miles. The cantilevered channel span has a length of 1,400 feet, a record in the United States that stood for more than twenty years, and provides a clearance of 185 feet above mean water level. This extraordinary project, designed and built under the supervision of C. H. Purcell, was the culmination of a long history of proposals going back to 1900. It is the most important of the San Francisco bridges because it ended the city's dependence on the ferry service that formerly united it with the extensive urban area on the east side of the bay.

The bridge at New Orleans (1933–36) sprang from an equally urgent need, but its site and its dual purpose posed a peculiar combination of problems in location, design, and construction. With the end of World War I the city began its rapid expansion as an ocean port and as a center of manufacturing and oil refining, activities that were to be even more powerfully stimulated by World War II and the post-war industrial boom. Like San Francisco, New Orleans is largely surrounded by water and depended on ferry service to reach the communities and rail lines on the opposite shore of the Mississippi. Citizens' organizations had begun to propose the building of a bridge as early as 1892, but the character of the region raised difficulties that were long insurmountable. The river seemed an impassable barrier—3,000 feet in width, 214 feet deep in the channel at Algiers Point, normally carrying 1,000,000 cubic feet of water per second, subject to destructive floods before the completion of Bonnet Carre spillway in 1936—impressive features that would call for heroic measures. The topographic and geological character of the area offered further discouragement. Since the land is scarcely above sea level, a bridge that would satisfy the Army Engineers' rule of a 135-foot minimum clearance would have to be built with extremely long approaches for rail lines. And from the standpoint of construction realities, there is no bedrock in the lower Mississippi valley; sand, clay, and alluvial sediments have accumulated to a depth variously estimated at 4,000 to 8,000 feet. By 1932, however, engineering surveys had established the feasibility of construction, and in the same year Governor Huey P. Long, who devoted his administration to a huge public works program, created the Public Belt Railroad Commission to issue the bonds necessary to finance the building of the bridge and a union terminal in New Orleans (the latter was not completed until 1954). Modjeski, Masters and Chase were appointed engineers for the project, and construction began in 1933.

The cantilever channel span of the Huey Long Bridge is 790 feet long, a modest length when compared with the main spans of many other bridges, but this represents only 3.5 per cent of the end-to-end length of 22,996 feet. The long railroad approaches, which must be held to a gentle grade for the economical operation of trains, make the structure the longest continuous steel bridge in the world. The double-track rail line lies between the subdivided Warren trusses of the river crossing, and the divided highway is carried outside of them on cantilevered brackets. This enormous load of steelwork, concrete, and moving traffic, astonishingly enough, rests on sand. Test borings by the engineering staff revealed a thick, compact layer of sand at a depth of 180 feet below sea level, which was sufficient to place the sand under a natural load of 14,000 pounds per square foot. The problem was to transmit the total load delivered by the pier to a bearing surface with an area great enough to hold the unit load below this figure. The solution was to place the pier on a distributing block resting in turn on a caisson in the form of an immense concrete box divided into cells by longitudinal and transverse walls. The whole structure of caisson, distributor, and granite-faced pier footing was built up by the sand-island method, first used in the construction of the Carquinez Strait bridge of the Southern Pacific Railroad (1928–30). In this heroic procedure a huge circular cofferdam is first sunk to a weighted willow mattress on the river bed. The water is pumped out of the enclosure and the open volume filled with sand. The caisson is then simultaneously sunk and built up as dredging proceeds through the sand and into the river bed. This costly method is used to keep the caisson level and steady in highly fluid sediments and to avoid pneumatic excavation at great depths. The technique was adopted by the railroad at Carquinez Strait because of the trouble with caisson tipping during the construction of the earlier bridge over the same waterway (see pp. 219–20).

The second bridge built at New Orleans, a single-deck highway span, is a cantilever structure of pure Warren trusses also designed by Modjeski and Masters (1955–58). The main span of 1,575 feet makes it the longest trussed span in the United States, but its now thoroughly standardized form of two anchors, two cantilevers, and suspended structure has become commonplace. Even though this makes for uninteresting bridges, it indicates the extent to which the whole process has been rationalized. Once the span and the clearance have been fixed by the characteristics of the waterway and its traffic, standard highway loadings and construction in Warren or Pratt trusses admit of little variation.

Two truss forms reappeared in American bridge construction in the twentieth century, although both had been invented before 1900. The older with respect to origin is the K-truss, developed by Stephen H. Long for the bottom frame of the truss bridge he patented in 1830 (see p. 59). Its chief virtue is that the two short diagonals, meeting at the center point of the transverse member, are less prone to buckling than the long diagonal extending across the full panel (Fig. 79). The K-truss seems to have disappeared after Long's death, to be revived probably as a new invention by Ralph Modjeski for the main supporting trusses of the tremendous cantilever bridge that he built over the St. Lawrence River at Quebec (1911–17). The extreme truss depth for the 1,800-foot main span made it imperative that the diagonals be held to the shortest possible length. The K-truss was reintroduced into the United States before the Quebec bridge was completed and reached its greatest length a number of years later in the channel span of the highway bridge over the Atchafalaya River at Morgan City, Louisiana (Fig. 79). Designed by N. E. Lant and built in 1933, the structure includes a river crossing of 608 feet. The K-truss is frequently employed as wind-bracing, exactly as Long proposed, and is especially common in steel arch bridges, but it has also continued to play a modest role as a main supporting structure.

The more recent of the two innovations is radically different from any of the triangulated forms. The rigid-frame truss was invented in 1896 by the Belgian engineer Arthur Vierendeel for short-span bridges (Fig. 80). There is reason to believe, however, that it was independently developed in 1900 by the American engineers Mason R. Strong and Octave Chanute for the bents of the Kinzua Creek Viaduct of the Erie Railroad. The chief characteristic of the Vierendeel truss is the absence of diagonals, rigidity being obtained by rigidly connecting the posts to the chords by means of fillets. The deflection under load that would be distributed as direct tension and compression in the web members of the triangulated truss produces a complex pattern of bending in the verticals of the Vierendeel form, resulting in stresses that reach a high concentration at the joints. The chief advantage of the truss is a matter of economy: the number of web members is one less than half that needed for the simplest of conventional forms, and in a truss with parallel chords, these are uniform in length.

The peculiar design of the bents supporting the Erie viaduct had no influence on American bridge building, and Vierendeel's invention does not seem to have been adopted in the United States for more than forty years after its initial appearance. In 1937 the Corps of Engineers began to use Vierendeel trusses in street bridges over flood-control channels in Los Angeles, which were designed by L. T. Evans (Fig. 80). But the truss has a limited use and has never become popular for the primary structure of truss bridges in the United States. On the other hand, a special application of the form has given it a valuable role in the structural system of suspension bridges. Since the steel towers of such bridges act partly as vertical cantilevers which must be braced against wind loads and cable sway, the necessary rigidity can be obtained either through diagonal bracing or rigid-frame construction. If the tower includes more than one frame, it is in effect a vertical Vierendeel truss, although closer in form to the bents of the Kinzua Creek bridge than to the Belgian invention. The first Vierendeel towers were designed by Steinman and Robinson for the Waldo-Hancock Bridge, built between 1929 and 1930 over the Penobscot River at Bucksport, Maine.

The visual prominence of large truss bridges obscures the fact that the overwhelming majority of American bridges are small girder spans of steel or concrete, with the metal predominating for railroad structures. The girder has had a continuous history in the United States since 1846, and the built-up flanged type, which had been developed by the time of the Civil War, has retained a stable form since its inception. The universal employment of ferrous metals was challenged in 1933, when the floor frame of the Smithfield Street bridge in Pittsburgh was rebuilt in aluminum, and again in 1946, when the Aluminum Corporation of America built an aluminum girder span to carry a spur track to its plant at Massena, New York, but the original promise was never realized and the metal has been extremely rare as a structural material. Although the simple through girder has remained standard for railroads, the continuous form became common for rapid transit and highway bridges shortly after the turn of the century. Continuity is generally revealed in the lower profile of the girder, which drops down at the supports to provide additional depth at the region of maximum bending. The long horizontal lines, the slender proportions, and the smooth surfaces provided by welded connections, which were introduced into bridge construction by the Westinghouse company in 1928, often give the commonplace girder an elegance and purity of form that is rare among truss bridges. Since the basic character admits of little variation, progress is reflected almost entirely in steady lengthening of span. The present American record for conventional flanged girders is the 450-foot main span of the bridge carrying Interstate Highway 210 over the Calcasieu River at Lake Charles, Louisiana (1962–63). The designing engineers, Howard, Needles, Tammen and Bergendoff, have a long roll of distinguished bridges to their credit.

The rigid frame was adapted to girder-bridge construction shortly after it was introduced into the framing of buildings. It was first used in the remarkable system of floor beams in the Sciotoville bridge, although in this case the usual action is reversed since the frames are inverted. The horizontal member is deflected downward, away from the legs, which must then act like hangers and be subject chiefly to tension. However, the aim here is essentially the same as it is when the frames are built in the normal position: to distribute the stresses throughout the vertical and horizontal members so that all share to the maximum degree in sustaining the load. The first true rigid-frame girder bridge in the United States was constructed between 1928 and 1929 to carry the Bronx River Parkway over the tracks of the New York Central Railroad at Mount Pleasant, New York. The designer, Jay Downer, chief engineer of the Westchester County and Bronx

River Parkway commissions, deserves the major credit for making the rigid frame in both steel and concrete a common form in American bridge construction. The Mount Pleasant bridge consists of five parallel steel frames of 100-foot span, its naturally graceful and dynamic form hidden by a masonry cover. This passion for pseudo-rusticity is the only defect in the otherwise admirable program of highway building and scenic preservation in the Westchester County Conservation Plan.

The first rigid frame designed for railroad loading was built in 1930 by the Kentucky and Indiana Terminal Railroad to carry its tracks over a street in Louisville. Since railroads are seldom concerned with the appearance of a bridge, the sturdy form of the Louisville span stands undisguised, clearly expressing its scientifically calculated efficiency. The basic structural action can be read from the visible profile of this bridge: the slightly arched undersurface of the horizontal member indicates that the bending moment decreases from knee to mid-span; the enlarged depth at the knee reveals the high stress concentration in that region; and the downward-tapering legs tell us that the bending moment contracts to a minimum (theoretically zero) at the bearing points. The form was to be greatly refined in the multiplicity of overpasses that came with the urban expressway systems after World War II, but all the essential features are prefigured in the pioneer structures.

The great breadth of modern high-speed arteries and the heavy traffic loads imposed on them demanded a much stronger floor-framing system than was possible with the older frame of two girders joined by transverse deck beams. The common practice now is to pour the reinforced concrete deck slab on the top flanges of a dense array of girders that are braced by transverse beams welded to the primary members. This technique provides the necessary strength but does not achieve maximum efficiency because the deck functions only to distribute the traffic load over the girders and plays no role in the load-supporting action of the steelwork. A more organic solution to the problem is the structural technique known as orthotropic design, developed by German engineers and introduced into American practice in 1964 for the Poplar Street span over the

Mississippi River at St. Louis. The designing engineers of this pioneer structure were Sverdrup and Parcel. In this system the primary supports are two parallel continuous box girders with conventional bottom flanges. The girders carry a steel deck that forms an integral unit with them and acts as their common top flange, the three elements together forming a hollow steel box with an open bottom. In the action of the orthotropic bridge, the stresses induced by traffic and wind loads are distributed in a continuous pattern throughout the deck and the girders, which act together as a rigid unit in sharing the load. The chief advantage is that for a given quantity of metal the orthotropic girder offers greater resistance to torsional forces than the conventional form. Another type of girder bridge that came from Germany in the years following World War II is the cable-stayed girder. The earliest ancestry was the system of radiating stays that Roebling adopted for his suspension bridges, but a form closer to the modern variant was invented by Joseph M. Wilson and first used in 1874 to carry 40th Street in Philadelphia over the Pennsylvania Railroad tracks. In the stayed girder, introduced to the United States in 1969, the load is shared between the two plate girders at the sides of the deck and a series of cable stays radiating outward and downward to the panel points of the deck frame from the top of a steel bent much like a suspension bridge tower. Since the stays form rigid figures with the tower and girders, they also aid in resisting wind loads.

The steel arch bridge is a far larger and more impressive work than the commonplace girder; yet in spite of its size and the diversity of conditions it has been designed to meet, it has retained the most stable form of all modern structures. All the basic types of metal arch developed in the nineteenth century have been used simultaneously throughout the twentieth, with progress confined mainly to refinement and simplification of form accompanied by a steady increase in weight and length of span. The components of the arch bridge are few in number, admitting of little variation in their essential form and relationship: a pair of ribs, each surmounted by a stiffening truss in large bridges; wind- and sway-bracing set transversely between the ribs and between the top chords of the trusses; hangers or posts to support the deck; abutments with sloping faces to take the thrust of the arch. The basic fact in the action of the steel-rib arch is that the load is sustained entirely by compression in the rib; the heavy trusses which often lie above it act only to provide rigidity against bending and buckling induced by moving loads.

As in the case of the truss bridge, the big steel arch was dominated by the demands of the railroad until the mid-twenties, when the requirements of highway traffic became paramount. At the turn of the century the hinged arch had risen to a position of prominence, but it has never entirely superseded the fixed variety. The two-hinged form, with its greater rigidity, is generally the choice for railroad spans and the very largest highway spans, whereas the three-hinged has been progressively restricted to the lighter highway structures. The classic among three-hinged arches is the bridge of the Minneapolis, St. Paul and Sault Ste. Marie Railway (Soo Line) over the St. Croix River near New Richmond, Wisconsin. Built between 1910 and 1911, the structure was designed by Claude A. P. Turner, who had already received national attention for his revolutionary innovations in concrete construction (see pp. 243–44). It carries the track high above the valley in a straight line that replaces an old alignment of grades and sharp curves. The light traffic of the Soo Line and the smaller axle loading of its locomotives allowed Turner to reduce the arch spans to essentials of remarkable purity and delicacy—shallow deck girders, slender three-hinged arch ribs identical in span and rise, and an extremely light and open pattern of spandrel bracing. The central hinge is a curious sliding joint designed to lock into a rigid unit under the weight of the train and thus intermittently to transform the rib into a two-hinged arch. The 350-foot span of the individual arch was unusual at the time for hinged construction under railroad loading.

The perfect antithesis to the Soo Line bridge in size and mass is the great Hell Gate arch over the East River in New York City (Fig. 81). This masterpiece is the foremost creation of Gustav Lindenthal, who had forty years of bridge-building experience behind him when he undertook the commission. The Hell Gate span is a part of the grand scheme for the Pennsylvania Railroad's New York Extension, and comparable resources of talent and money were necessary to bring it to realization. The construction of the railroad's New York station brought the passengers into Manhattan but left unsolved two closely related problems. A sizable proportion of passenger traffic and a heavy volume of freight tonnage moved through New York to and from Montreal and the New England cities via connecting lines. After the opening of the station in 1910 passengers could transfer with only the normal frustrations to the New Haven Railroad at Grand Central Terminal, but freight traffic had to be moved by car float across the Hudson and up the East River from Jersey City to Port Morris in The Bronx. The opening in 1888 of a through rail route from New England to Pennsylvania and the Pocohontas region via the Poughkeepsie bridge placed the Pennsylvania Railroad at a disadvantage. To reduce the time and the cost of the long water transfer at New York, to improve the flow of traffic through the metropolitan gateway, and to maintain its own competitive position, the railroad had to find a way of closing the gap between its own line and the New Haven system. The extension of the line under the East River to Queens brought the two railroads close to a junction point, but to connect them raised the awesome spectacle of bridging Hell Gate. The momentous decision to implement the program was made in 1907, and construction began two years after the New York Extension was completed. The huge arch over the upper end of the East River was built between 1914 and 1916, and traffic began to move over the new connection in the following year.

Hell Gate, a relatively narrow channel lying southeast of Ward's Island, offered a particularly forbidding combination of the difficulties that John Roebling first contended with. The constricted waterway and the neighboring rock formations generate the swiftest and most turbulent tidal currents in the harbor area, with the result that construction by means of falsework was out of the question. The heavy river traffic demanded a single high-level span, the massive abutments of which would have to be founded securely on bedrock. Good bearing rock lies 70 feet below the surface, but at the site of the Ward's Island abutment it is split by a wide fault that would threaten the stability of any structure built above it. The first necessity, accordingly, was to cover this break, which was accomplished by pouring an underwater system of concrete arches and cantilevers over the opening. On completion of the abutments and the towers above them, the steel arch was erected by cantilevering the two halves out under backstays that were supported on temporary steel bents fixed to the tops of the towers. The two segments of the arch, brought into exact alignment by means of 2,500-ton hydraulic jacks, were closed by a hinged joint at the crown, which was riveted fast as soon as the stiffening truss was erected.

The primary supporting members of the bridge are the two parabolic ribs, hinged at the springing points, that constitute the arch proper and take the full compressive stress. The over-all span of the arch between abutment faces at the deck is 1,017 feet, and the clear span between hinges is 977 feet 6 inches, making it the longest arch in the world at the time. Although it can no longer claim this record, it is still the heaviest structure of its kind, having been designed to carry 24,000 pounds per lineal foot of live load and 52,000 pounds of dead load. The frame that supports the four-track railroad line is suspended below the arch by rigid hangers for most of its length but stands above it on posts near the abutments. The arched Pratt trusses above the ribs function entirely as stiffening elements designed to absorb bending forces induced by live and eccentric loads. Wind- and sway-bracing is located on the transverse line between the ribs and between the top chords of the trusses. Since there is no bending in the arch at the hinges, it would have been possible to carry the top chord down to the hinged end of the rib, but the recurved shape of this chord was adopted to provide adequate clearance under the transverse bracing for trains and the overhead electrical system. The arch rib and truss together give the appearance of a fixed arch, which is further suggested by the protective coverings of steel plate over the hinges. Although the masonry towers have no structural function above the skewbacks, they are appropriate to an arch of this size as a kind of visual containment or solid boundary for the ends of the stiffening truss.

The longest arch span in the world is that of the highway bridge over the New River gorge near Charleston, West Virginia, constructed in 1975–77 with a clear span of 1,700 feet. The record was long held by the highway bridge built between 1928 and 1931 to span Kill van Kull between Bayonne, New Jersey, and Port Richmond on Staten Island (Fig. 82). The third bridge of the Port of New York Authority (established in 1921), it was designed under the supervision of the Authority's chief engineer, Othmar H. Ammann, with Cass Gilbert as architect. The bridge is structurally similar to Hell Gate but considerably different in appearance. Indeed, in one respect it is markedly inferior, through no fault, as we shall see, of the engineering and architectural designers. The parabolic arch ribs span 1,675 feet between hinges for a rise of 266 feet, making a flatter and more dynamic curve than that of Hell Gate. The Pratt trusses above the ribs are the stiffening elements, and the arch is braced laterally by a K-truss frame. The curve of the top chord is again calculated for maximum truss depth at the ends to provide adequate clearance between the roadway and the transverse bracing. Where the deck lies above the arch, the framing is carried on columns of H-section; elsewhere it is hung by cable suspenders from the underside of the ribs. The traffic and dead loads of Bayonne Bridge are much lower than those of Hell Gate, but the extreme span of the newer structure results in the enormous thrust of 30,000,000 pounds per rib at each skewback. The abutments that take this load are solid concrete blocks whose upper surfaces rise a little above water level at high tide to provide a base for the steel-framed towers that carry the deck over the abutments.

It is in the treatment of these towers that the design of Bayonne Bridge went astray. Gilbert's original plan called for covering the steelwork with granite sheathing, but the Port Authority rejected his proposal on the ground of its high cost. The result was unfortunate in two ways: the framework is inharmonious with the appearance of the main structure and the approaches; further, since the top chord of the truss appears to be in compression, the frame in turn appears too flimsy to support it and is ambiguous in its suggestion of a contact. The adoption of Gilbert's proposal, which suggests mass and solidity, would have prevented these irritating weaknesses in the design.

Among the nineteenth-century inventions in iron bridge construction that have been used in the twentieth century is the tied arch, in which the thrust of the arch is sustained by tension in a horizontal tie extending between the springing points rather than by the mass of the abutment. Tied-arch bridges have been built wherever conditions are such that the conventional mode of construction is either inapplicable or prohibitively expensive. The first tied arch designed for a broad navigable waterway is the West End–North Side Bridge over the Ohio River at Pittsburgh. It was built between 1930 and

1932 from the design of V. R. Covell, for many years chief engineer of Allegheny County and the author of many of the great number and diversity of spans in the Pittsburgh area. The complication in the case of the West End arch was that the abutment pier standing in the water near the south bank had to be held to a 25-foot depth in the line of the roadway to maintain an adequate width of channel and to offer a minimum obstruction to the flow of water; but the necessary vertical clearance at this pier would have resulted in an overturning moment from the arch thrust too great for the narrow, isolated block to sustain. In the arch that Covell designed, the tie consists of eight parallel lines of I-beams located below the deck and extending the full span of 780 feet. The bridge was the longest tied arch until 1943, when the Julien Dubuque Bridge over the Mississippi at Dubuque, Iowa, was completed.

Although the fixed arch is the oldest form by virtue of its masonry precedent, it continues to lead a vigorous life in bridge construction. A particularly handsome work of this type is the highway bridge that crosses the Niagara River between Queenston, Ontario, and Lewiston, New York. Designed by the engineering firm of Hardesty and Hanover and constructed between 1960 and 1962, the bridge stands at the edge of the Niagara escarpment, where the expanded width of the gorge required an arch of 1,000-foot span. The extreme refinement of its form eloquently expresses the rigor of modern steel construction. Here the elements are reduced to minimum essentials: two fixed ribs of smooth plate and uniform depth, spandrel posts without diagonal bracing, K-truss bracing between the ribs, and pyramidal skewbacks set with their axes in the line of the rib tangents. Only in concrete would it be possible to secure a greater purity and homogeneity.

As in the other forms of the steel bridge, the fundamental characteristics of the suspension bridge had been developed by 1900. One can see this clearly in the first big span of the new century, although it seems somewhat clumsy compared to later designs. Manhattan Bridge (1901–09), the fourth to cross the East River in New York, has a main span of 1,470 feet, somewhat shorter than that of the earlier Brooklyn Bridge (Fig. 83). O. F. Nichols was the chief engineer of the project, and Carrère and Hastings were the architects. The structure is steel throughout: the towers are in effect trussed rigid frames in which the horizontal member has an arched bottom chord, and the stiffening trusses are in the Warren form. Except for the towers, which were ordinarily built with diagonal bracing rather than as single frames, this structural system remained substantially fixed for nearly twenty years after the completion of the Manhattan span.

Three important innovations appeared in close succession during the decade of the twenties. The first was the self-anchoring suspension bridge, introduced into American practice by V. R. Covell in his designs for the Sixth, Seventh, and Ninth Street bridges over the Allegheny River at Pittsburgh (1926–28). The self-anchoring span is the suspension equivalent of the tied arch, although the action is exactly opposite: the load on the cables is transmitted to longitudinal girders under the deck of the bridge, but since the cable is in tension, these girders must necessarily be in compression. In the case of the three Pittsburgh bridges, the technique was adopted because of insufficient space for anchor blocks in the densely built areas around the approaches. Another unusual feature of Covell's bridges is the presence of stiff eyebar chains in place of wire cables. Chains of this kind are composed of bars with enlarged ends pierced to receive the pins that join one bar to the next. Although a number of such bridges were built in the late nineteenth century, the form had disappeared until Covell briefly revived it at Pittsburgh. The other two inventions are both embodied in the Waldo-Hancock Bridge (1929–30) of David B. Steinman and Holton Robinson: one was the application of the Vierendeel truss to the tower construction, and the other the use of prestressed wire-rope cable. The towers contain no diagonal bracing, rigidity being secured by means of shallow transverse trusses riveted throughout their depth to the posts. The cable was prestressed after fabrication on the ground in order to give it its maximum elongation under tension and thus to avoid further stretching under the traffic load.

Much of the development of the suspension bridge in the first quarter of the present century constitutes preparation for the three great structures that were completed in the decade of the thirties—George Washington, San Francisco Bay, and Golden Gate. Although the span lengths of all three have now been exceeded by that of Verrazano-Narrows Bridge at New York, they remain the classics, and George Washington is still the greatest suspension bridge in capacity and total weight (Fig. 84). The history of proposals for a bridge over the Hudson at New York goes back to 1868, when John Roebling first suggested the feasibility of its construction. The possibility of successful action came in 1910, when the governors of New York and New Jersey appointed the Interstate Bridge and Tunnel Commission to investigate the matter. It was this commission that recommended the Fort Washington site, where high palisades and good bearing rock offered the greatest natural advantages to economical construction. The establishment of the Port of New York Authority in 1921, which was charged with planning and financing bridges and tunnels, brought the whole project close to realization, although it was six years before the Authority issued the bonds and the engineers completed their plans. The bridge was placed under construction in 1927 and opened to traffic in its original form on October 25, 1931. The designing team included Othmar Ammann as chief engineer, Leon S. Moisseiff and Allston Dana as consultants, and Cass Gilbert as architect. The 3,500-foot main span was at the time nearly double the length of its nearest competitor.

The original plan of the bridge called for an upper deck for vehicular traffic and a lower deck for a four-track rapid transit line, but these plans were eventually changed in favor of a second vehicular deck, which was added between 1961 and 1962. The provision for a rapid transit deck as well as the long span gave George Washington Bridge its great weight and over-all size. The towers, space frames of unparalleled size, stand 635 feet above the top of the piers, and the four 36-inch cables contain 107,000 miles of wire and originally sustained a total load of 200,000,000 pounds. The sheer mass of the bridge gave it sufficient stability against wind to make stiffening trusses unnecessary until the addition of the lower deck required them as part of the supporting framework as well as for wind-bracing. The New York anchorage is the conventional concrete monolith, but the one on the New Jersey side is unique. Two sloping funnel-shaped tunnels were blasted out of the igneous rock of the palisade, and the cable was anchored in the rock itself rather than in concrete.

The treatment of the towers provoked a controversy that paralleled the question of the abutment towers for the Kill van Kull bridge. Gilbert originally proposed that the naked framework of the huge towers be covered with granite sheathing. His intention was practical as well as aesthetic: the protective covering would preserve the steel from corrosion and provide additional strength for supporting the cable load. But the Port Authority discovered that it would cost less to paint the towers regularly than to build the masonry envelope and that the small increase in strength derived from the masonry would be unnecessary, even with the addition of the rapid transit deck. Once again Gilbert lost the argument, but in this case there was something to be said for both sides. The intricate pattern of the steelwork in the towers is confusing to the uninitiated and inharmonious with the long straight lines and parabolic curves of the suspended structure. A simpler construction with extensive plane surfaces, as in the towers at Golden Gate, would have given the whole work a more unified appearance. On the other hand, the open pattern of the frame is more appropriate to the buoyancy of the suspension bridge than a solid masonry covering, which suggests great weight and earthbound mass. And finally, one may argue that the spectacle of a space frame the size of the George Washington towers is an experience one ought to have, for it offers something that far transcends the ordinary.

The simultaneous construction between 1933 and 1937 of the two suspension bridges at San Francisco was a building feat comparable to the erection of Brooklyn and Eads bridges sixty years earlier. There is a marked difference between the two in structural character, over-all size, and appearance. The western half of San Francisco Bay Bridge is unique among structures of its type, for it is made up of two suspension bridges placed end-to-end with a common anchor pier at the center of the crossing (Fig. 85). The main spans of both of the bridges are 2,310 feet in length, well below that of George Washington, but the total length of deck under cable is a record 9,260 feet. The common anchorage is a solid concrete block with the dimensions of a skyscraper: its over-all height is 514 feet, of which 220 feet are below water and 294 feet above. The four towers are diagonally braced frames rather than Vierendeel trusses. Its great size, together with the long reach of open water below it, make the Bay bridge an impressive spectacle, although its appearance suffers a little in comparison with its neighbor across the Bay—the Golden Gate. The heavy diagonals of the towers do not harmonize with the dominant curves and planes of the suspended structure;

indeed, they seem a little crude and old-fashioned compared with the aerial delicacy of cables and deck. The common anchor pier is even more discordant, although its presence is necessary because a continuous three-span bridge would be too flexible for automotive and rapid transit traffic. This huge block of concrete is out of scale in dimensions and mass with the rest of the structure, and its prominent location at the center makes it the most conspicuous element of the bridge. For the sheer magnitude of the work, however, we can have nothing but admiration for the designing staff, which included C. H. Purcell as chief engineer and Glenn B. Woodruff, Ralph Modjeski, Leon S. Moisseiff, and Charles Derleth, Jr., as consultants.

Golden Gate Bridge, though smaller than the Bay crossing in over-all size, offered a greater challenge to the builder by virtue of its site (Fig. 86). The one-mile width of the waterway had to be crossed by a single span at high level, which would stand unprotected before the fury of Pacific storms. The first to take up the challenge was Joseph B. Strauss, who eventually became chief engineer of the project. An authority on movable bridges and a former assistant to Ralph Modjeski, Strauss made his first proposal in 1920, for a combination of suspension and cantilever bridges. Doubts about the feasibility of the whole idea, difficulties of financing the work, and economic depression delayed the sale of bonds and the start of construction until 1933. The most potent factor in bringing the plan to realization was the tremendous increase in automobile traffic, which intensified all the problems arising from San Francisco's isolation. Strauss assembled a brilliant team of designers and consultants to complete the plans, among them O. H. Ammann, Moiseiff, and Derleth.

The main span of Golden Gate Bridge, which extends 4,200 feet between the towers, was the longest in the world for nearly thirty years; the tower height of 746 feet above mean low water remains unsurpassed. Much of the formal excellence of the bridge is derived from the towers, which are unique both in structural character and appearance. They are Vierendeel trusses composed internally of a great multitude of small cubical cells bounded by steel plate, a form of construction adopted to secure minimum weight with maximum strength and rigidity against earthquake shocks. The outer surfaces are treated as narrow bands of steel plate and are scored with shallow inflectional lines to heighten the sense of verticality. The extraordinary setting of the bridge intensifies its visual power: the open ocean lies to the west, paralleling the long deck line; the south tower stands in extreme contrast to the modest scale of San Francisco's streets, and the north tower nicely measures the blocklike masses of the Marin County mountains.

A century of progress in the development of wire-cable suspension bridges was violently interrupted by the destruction of Tacoma Narrows Bridge on November 7, 1940. The structure was designed to sustain the force of a 122-mile-an-hour wind, but it went down after four hours of punishment from a moderate 42-mile-an-hour gale. The bridge was regarded as aerodynamically stable, although questions had been raised about one feature of its design from the time it was announced. The standard practice for calculating the depth of stiffening trusses dictated a minimum truss depth of 31 feet for the Tacoma span. The designers—Lacey V. Murrow, chief engineer, and Leon Moiseiff, consultant—chose instead to use plate girders with only an 8-foot depth, which were thought to be adequate to maintain the rigidity of the deck on the basis of both tests and calculations. What actually occurred under the force of the wind was wholly unforeseen. After three hours of minor vertical oscillation, the deck suddenly began to twist rhythmically around the center line of the roadway, the twisting waves growing in amplitude up to 28 feet, at which point the ruined deck tore loose from the suspenders and fell into the water. There can be no question that this torsional action arose from radical aerodynamic instability, which was in turn a consequence of the extreme flexibility of the ribbon-like girders. When the Tacoma Narrows Bridge was replaced, the engineers used Warren stiffening trusses 33 feet deep.

With the end of World War II and another explosive rise in automobile traffic, the suspension bridge entered a new period of growth. The most dramatic achievement was the bridging of Mackinac Strait after seventy years of fruitless discussion and surveying. The first proposal to span the broad waterway that joins Lake Michigan and Lake Huron was offered in 1884; the construction of the bridge that turned these visions into reality began in 1954 and was completed three years later. The chief engineer of the project, David B. Steinman, had to contend not only with the usual hazards of wind and water but with snow and ice on a terrifying scale. Three times the winter storms reached such violence that construction had to be suspended. The heavy structural system, with its massive Vierendeel towers and deep stiffening trusses, bears witness to the wind and snow loads the bridge must resist. The main span of 3,800 feet places it second to Golden Gate Bridge, but the extremely long anchor spans give it a record length under continuous cable of 8,614 feet. The long deck-truss approaches, which give the whole structure an end-to-end length of more than four miles, heighten the sense of a particularly bold conquest of open water.

The span length of Golden Gate was finally exceeded with the construction, 1959–64, of Verrazano-Narrows Bridge between Brooklyn and Staten Island (Fig. 87). Designed by the engineering firm of Othmar Ammann and Charles S. Whitney, the bridge takes its name from Giovanni da Verrazano, the Florentine navigator who in 1524 became the first European to sail into the waters of New York Harbor. Sheer size and cost distinguish this structure, certainly impressive in its own right, yet somehow lacking the drama of the achievements at San Francisco and Mackinac. The main span is 4,260 feet long and stands with a clearance of 216 feet at mean water level. The towers are the simplest possible construction: they are plate-covered rigid frames in which the single transverse member above the deck has an arched undersurface—like two enormous narrrow gateways of smooth steel. The staggering cost of the Verrazano span—well over $300,000,000—indicates that the greatest challenge for the future engineer is likely to be economic more than technical. The near prospect of spending half a billion dollars on a single bridge discourages even the highway authorities, who have always consumed money and land with reckless abandon. For this reason the new metropolitan transit system of San Francisco passes under the bay in a tunnel rather than over it on a bridge. And so for the next development in building techniques we are compelled to return to the solution that the Pennsylvania Railroad provided in New York at the turn of the century.

Chapter Eighteen | Concrete Framing for Buildings

The first revolution in the building arts followed the introduction of iron as a structural material, and the second came about a century and half later with the application of concrete to all structural techniques, many of which would have been impossible without it. Its virtues are so numerous that it has far outdistanced any competitor: in 1963 the American building industry mixed 500,000,000 tons of concrete, whereas the metal industry produced 100,000,000 tons of steel, of which a substantial proportion went to non-structural uses. The chief advantages of concrete are its compressive strength, durability, low cost, and plasticity (that is, the ease with which it can be poured into any structurally viable form). It was long considered unattractive for high-quality exterior finishes, and rightly so, but the development of high-density concretes mixed with special aggregates has made the material competitive in appearance with the common forms of masonry. The major structural innovations of the twentieth century have been the products of concrete technology, and many of these have led to radical changes in the form and action of structural systems. No building material was treated to a more scientific investigation than concrete, with the consequence that by mid-century its chemistry, its internal structure, and its behavior under every condition are as well understood as the properties of familiar metals. Indeed, the engineers regard it as the most scientific material, one that allows the closest approach to the organic ideal, in which structural form exactly corresponds to the pattern of internal stresses.

But reinforced concrete, like all composite things, can never reach the ideal of perfect homogeneity—of an isotropic substance, as the engineers would call it. The material has certain inherent defects, chiefly permeability and capillarity, which allow moisture to penetrate it, subjecting the reinforcing to corrosion and causing the concrete to crack under freezing temperatures. Cracking as a result of thermal expansion and contraction, shrinkage during setting, slow deformation under prolonged load (creep)—all produce internal weaknesses and secondary stresses that must be foreseen in the initial calculations. Some of these characteristics were known as early as 1900, and others were later discoveries of French and German engineers; but even when they were known, builders were often at a loss to overcome them. Extraordinary structural achievements in concrete came early in the century, but careless design and inadequate building codes meant that progress was sometimes interrupted by disaster.

By 1910 concrete construction had become extremely popular in the United States: it was an exciting novelty, and enthusiasm for it paralleled the earlier passion for cast iron. The leading engineers and builders at the turn of the century were Ernest Ransome of San Francisco, Julius Kahn of Detroit, Schmidt, Garden and Martin in Chicago, and the various branches of the Ferro-Concrete Construction Company. The reinforcing techniques they used were variations on those developed by Hyatt, Ransome, Wayss, and Hennebique in the previous century: bars were densely set in zones of tension, and various shapes of bent and diagonal bars and J-shaped pieces known as stirrups were placed in the regions of shearing stress. These techniques were satisfactory enough with high safety factors, and they prompted bold experiments.

The first skyscraper of reinforced concrete came at the beginning of the century with the sixteen-story Ingalls (Transit) Building in Cincinnati (1902–3). The architects were Elzner and Anderson, and the designers and builders of its column-and-girder frame were the staff of the Ferro-Concrete Construction Company. The reinforcing system was based on Ransome's patents for fully developed reinforcing against tension and shear. The Kahn system was used in the frame of the fifteen-story Marlborough Hotel in Atlantic City, New Jersey, built between 1905 and 1906 with Price and McLanahan as the architects and Kahn's Trussed Steel Concrete Company as the structural engineers. This was the largest concrete building in the world at the time, and for all its fashionable attractions, a good part of the concrete is exposed in the exterior elevations. The floor rested on small densely spaced concrete beams, each containing a single "trussed bar" of the kind patented by the Kahn firm, which combined the longitudinal tension bar and the diagonal shear bars in a single unit. Other horizontal members ranged in size from the moderate wind beams to the huge distributing girders designed to carry offset columns under multistory loads. A more sober work for industrial purposes but with similar reinforcing is the factory of the Brown-Lipe Chapin Company at Syracuse, New York, built in 1908 as one of Albert Kahn's early essays in concrete architecture (Fig. 88). These buildings helped to fix the form of standard concrete framing up to the present time, to the extent that later variations were largely concerned with the more scientific distribution of reinforcing and the improvement in the quality of concrete.

In the development of wide-span enclosures the most important pioneer work is Terminal Station in Atlanta, Georgia, designed and built between 1903 and 1904 by the Baltimore Ferro-Concrete Company. The entire station except for the former train shed was built of concrete—piling, footings, columns, floors, roof, roof trusses, and the platform canopies beyond the open end of the train shed. This remarkable work contains a number of bold innovations that were not to appear as common features of concrete building for fifty years after its completion. The office space and the main waiting room are column-free areas, and the gabled roof slabs above them are carried from wall to wall by triangular Warren trusses, of which the largest spans 60 feet. The roof over the trapezoidal area between the station building and the train concourse rests on rigid-frame trusses in which the web system consists only of vertical members. Rigid-frame trusses in the bowstring form support the entrance drive over railroad tracks and driveways below grade level. The transverse skylights in the sawtooth roof over the concourse are carried along their lower edges by V-shaped troughs of concrete that are close to folded-plate construction (see pp. 278–80); the flanks of the monitors for the longitudinal skylights are built up of thin precast panels.

The special problem of wide-span framing for theaters in a region of earthquakes quickly led to a high stage of development in Los Angeles. The pioneer work is the Majestic Theater (1907–8), designed by the architects Edelmann and Barnett and the engineers Mayberry and Parker. Here the problem was unusually complex, for the building embraces a two-balcony theater and an eight-story office block, the upper three stories of which extend over the theater. In order to avoid columns in the theater the engineers placed the

84. George Washington Bridge, Hudson River, New York City, 1927–31. Othmar H. Ammann, chief engineer. Although several suspension bridges exceed George Washington in length of span, it is the heaviest bridge of its kind and, as a consequence, did not require stiffening trusses until the lower deck was added.

85. San Francisco–Oakland Bay Bridge, Calif., 1933–37. C. H. Purcell, chief engineer. The west half of this immense bridge complex is composed of two suspension bridges that meet at a common anchorage near the center of the waterway.

86. Golden Gate Bridge, San Francisco, Calif., 1933–37. Joseph B. Strauss, chief engineer. The great length of the main span, the setting, and the simplicity and unity of the design combine to make Golden Gate the classic of modern suspension bridges.

87. Verrazano-Narrows Bridge, New York City, 1959–64. Ammann and Whitney, engineers. The main span of Verrazano, 4,260 feet between the towers, is now the longest in the world for any kind of bridge.

88. Brown-Lipe-Chapin Company factory, Syracuse, N. Y., 1908. Albert Kahn, Inc., architects and engineers. Although the particular system of reinforcing has been superseded by more efficient types, the essential features of modern reinforced concrete framing were present in the Syracuse factory.

89. *Top:* Lindeke-Warner Building, St. Paul, Minn., 1908–9. C. A. P. Turner, engineer. Interior view. *Bottom:* Slab reinforcing in the Turner system. The flat-slab system invented by Turner greatly increased the economy of concrete construction by eliminating the girders and beams.

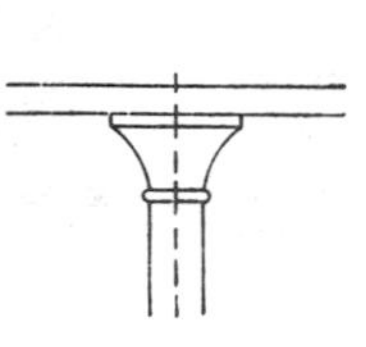

90. Brunswick Building, Chicago, Ill., 1963–65. Skidmore, Owings and Merrill, architects and engineers. The Brunswick is the largest and handsomest building in which all vertical and horizontal loads are divided between a solid core of shear walls and external bearing walls in the form of concrete frameworks.

91. Tunkhannock Creek Viaduct, Lackawanna Railroad, Nicholson, Pa., 1911–15. George J. Ray, chief engineer. The huge bridge possesses an undeniable power, but its Roman arches of concrete are thoroughly traditional in form and structural action.

92. Westinghouse Memorial Bridge, Turtle Creek, Pittsburgh, Pa., 1930–31. V. R. Covell, chief engineer. The central span of Westinghouse is still the longest concrete arch in the United States, a fact that reminds us, more than anything else, of the extent to which European engineers have progressed beyond American practice in concrete bridge design.

91

92

93. Highway bridge, Bixby Creek, Carmel, Calif., 1931–33. F. W. Panhorst and C. H. Purcell, engineers. The slender proportions and the setting make this standard parabolic arch a classic among American concrete bridges.

94. Western Hills Viaduct, Cincinnati, Ohio, 1930–32. Cincinnati Department of Public Works, engineers. A large-scale straightforward work of concrete girder construction.

93

94

95. Walnut Lane bridge, Lincoln Drive, Philadelphia, Pa., 1949–50. Philadelphia Department of Public Works, engineers. The first prestressed-girder bridge in the United States.

96. Gallipolis Dam, Ohio River, Gallipolis, Ohio, 1936–37. United States Army, Corps of Engineers, engineers. A typical roller crest dam, expressly designed for broad, relatively shallow waterways.

97. Stony Gorge Dam, Stony Creek, near Orland, Calif., 1926–28. Bureau of Reclamation, engineers. Stony Gorge clearly reveals the highly articulated character of the multiple-buttress dam.

98. Coolidge Dam, Gila River, near Globe, Ariz., 1926–28. Bureau of Reclamation, engineers. *Top:* Upstream face. *Bottom:* Downstream face. The only multiple-dome dam in the United States.

99. Bartlett Dam, Verde River, near Phoenix, Ariz., 1936–39. Bureau of Reclamation, engineers. Bartlett is a multiple-arch dam, which is a series of cylindrical vaults sustained by slab-like triangular buttresses.

100. Hoover Dam, Colorado River, near Las Vegas, Nev., 1931–36. Bureau of Reclamation, engineers. The highest in the world at the time of construction, Hoover is a spectacular example of the arch-gravity dam.

101. Grand Coulee Dam, Columbia River, near Spokane, Wash., 1933–43. Bureau of Reclamation, engineers. Grand Coulee is a straight-gravity dam, the largest ever built other than earth-fill dams.

102. Glen Canyon Dam, Colorado River, below Arizona-Utah boundary, 1956–66. Bureau of Reclamation, engineers. The dam and powerhouse during construction. Glen Canyon is a pure arch dam, that is, one that uses arch action alone to oppose the force of impounded water.

101

102

stepped slabs of the balconies on cantilevers anchored in the rear wall and extending outward on a sloping line for a maximum length of 30 feet. To support the office floors they employed rigid-frame trusses with rectangular web panels, each truss spanning 71 feet clear. All these members are heavily reinforced near the upper and lower surfaces against tension and rupture from buckling and in diagonal planes to take the shearing stresses.

The main tendency in American concrete construction in the period 1910–30 was to meet increasing loads simply by increasing the size of the structural elements. The result, which was on an elephantine scale, especially in buildings with large open interiors, revealed the diversity of requirements that could be satisfied by reinforced concrete but also the amount of valuable space that was consumed by the sheer quantity of material in long beams and trusses. For modest bay spans in multistory buildings a considerable saving in material and overhead space could be effected by transferring the action of girders and beams to the floor slab and thus eliminating the horizontal members. In flat-slab framing, as it is called, the floor slab rests directly on the columns and behaves somewhat like a continuous beam, bending down like an umbrella around the columns and like a saucer in the intermediate areas. Flat-slab construction was invented in 1900 by the Swiss engineer Robert Maillart and then independently developed in the United States by Orlando W. Norcross in 1902 and Claude A. P. Turner of Minneapolis, who first used it for the Johnson-Bovey Building in that city (1905–6) and obtained a patent on it in 1908 (Fig. 89). The distinguishing feature of the flat-slab frame is the flaring column capital, or mushroom capital, a conical spreading out of the cross-sectional area to reduce the concentration of shearing stress around the circular disc where the slab meets the column.

The economies of flat-slab construction were most obvious in the case of warehouses and freight-handling facilities, where overhead space is at a premium. The huge warehouses of the United States Army at Brooklyn, New York, provided the most convincing demonstration of its soundness, for these were the largest and most heavily loaded concrete buildings in the world at the time of their construction, 1917–19. They were designed under the direction of George

W. Goethals, chief engineer of the Panama Canal, with Cass Gilbert as architect. Equally impressive for the variety of its structural elements as well as its size is the Starrett-Lehigh Building in New York (1929–31), a nineteen-story structure that provides manufacturing, freight handling, and warehousing facilities for a variety of small businesses. The engineers were Purdy and Henderson, and the architects R. G. and W. M. Cory and Yasuo Matsui. The floor slabs are carried on steel columns up to the third floor and on concrete mushroom columns above that level, except in the area of transition from the wide spans of the driveways to the ruling bay span in the rest of the building, where the floors are supported by concrete Warren trusses with a maximum span of 40 feet. The slabs are cantilevered beyond the peripheral columns so that the curtain walls can be opened to continuous windows on all elevations. A novel planning feature of the Starrett building is the "vertical street" in the central area, which is a hollow core containing truck elevators, electrical conduits, and pipes for water, sewage, steam, and gas.

Subsequent progress in reinforcing techniques and the chemistry of concrete made it possible to eliminate the mushroom capital and the thickening of the slab around the column in the Turner system. Flat-slab framing eventually reached the point of maximum simplicity, marked by slabs of uniform depth and columns of uniform section, which freed the vertical space entirely of beams, capitals, and other structural details. This is best demonstrated in the Chicago hotel known as Executive House, built between 1956 and 1958 after the designs of the architect Milton Schwartz and the Miller Engineering Company. The thirty-nine stories and penthouse of this building rise to only 371 feet above grade, giving a floor-to-floor height of 8 feet 10½ inches for the standard ceiling height of 8 feet. The floor slabs rest on a combination of columns of rectangular section and four interior solid walls that rise continuously through the height of the building—shear walls, as they are called, now common in high buildings carried by steel as well as concrete framing. They serve a two-fold purpose, to support part of the floor and roof load and to provide rigidity against wind loads. In Executive House, additional wind-bracing exists in the form of two rigid slablike bents extending from the first to the sixth floor in the end walls. The caisson wells were sunk to bedrock, 118 feet below grade, by means of the Benoto caisson drill, which marked the first use of the French invention in American practice. The mechanization of this tedious process reduced the time of drilling and concreting caissons from nearly a year to 90 days.

The flat-slab frame suggested the possibility of using the slab as a vertical bearing member, eliminating the column and maintaining a uniform geometry of construction throughout the building. Vertical slabs were introduced into bridges by Robert Maillart in 1900 but were regarded as unacceptable by others because of the belief that the thin slab, like a narrow column, would buckle under load. Experiments carried on in Europe around 1930, however, indicated that this was not the case, for the reason that contiguous vertical elements of the slab tend to buckle in opposite directions, the reverse deflections thus canceling each other. Construction in vertical and horizontal slabs is called a box frame, although it is not true framing, which is composed of separate beams and columns. Since the reinforced concrete box is a rigid figure, the box frame is a rigid structure that requires no additional bracing. The great defect of the system is that interior partitions are fixed structural elements no one of which can be removed. For this reason box frames are rare, but they offer a considerable economic advantage in the uniformity of the formwork.

The largest work of box framing in the United States, and possibly the only one other than hollow-box foundations, is the Stateway Gardens project of the Chicago Housing Authority, constructed between 1955 and 1958 from the plans of the architectural and engineering firm of Holabird and Root. The entire group embraces two ten-story and six seventeen-story buildings, all of which are supported throughout by a box-frame system with 7-inch floor slabs and 8-inch wall slabs. The thickness of the vertical elements was enlarged slightly to withstand wind loads and to increase the resistance to buckling. The frames of these high narrow blocks presented a novel and arresting appearance, rare in American building, before they were largely hidden by the unimaginative treatment of the brick curtain walls common to public housing.

Perhaps the most far-reaching innovation in concrete framing is the construction of exterior walls as load-bearing frameworks of columns and beams. The load-bearing screen wall (or window-truss wall, as it is sometimes erroneously known) was developed in 1959 by the architectural firm of I. M. Pei Associates, working in collaboration with the engineer August E. Komendant. It was given its first practical demonstration in two large apartment groups built simultaneously between 1959 and 1961, the Kips Bay Plaza Apartments in New York and the University Apartments in Chicago. The invention grew out of a complex of economic, structural, and aesthetic motives. In an apartment building the irregular disposition of interior partitions arising from variations in room size ordinarily requires an irregular pattern of window sizes and window spacing in the elevations of the building. In order to maintain disciplined column spacing (as opposed to the "scatter columns" common in flat-slab construction), the architects decided to split the standard peripheral columns into very narrow, equally spaced members, four to each bay in the case of the New York and Chicago apartments. These light columns were then joined into an integral pattern by horizontal members located at the levels of the floors. The chief economic advantage is that the small identical elements of the wall frame make possible the repetitive use of identical, easily handled forms. The transformation of the walls into rigid structural frames allows them to transmit wind loads as well as the outer floor and roof loads directly through peripheral girders to the heavier columns at the base of the building. With the later addition of shear-wall construction, all loads could be divided between the solid interior walls and the framed exterior walls, thus providing another means of eliminating interior columns. The high quality of modern concrete, combined with careful workmanship in pouring the material and in stripping the forms, produces a finished surface fine enough in texture and color to make a masonry veneer unnecessary. The bearing screen wall not only expresses directly the strength and durability of concrete construction but combines to an almost paradoxical degree the qualities of lightness and openness with depth, mass, and vigorous pattern.

The inherent power of such a wall expresses itself most effectively in the Brunswick Building in Chicago, a thirty-six-story tower designed by Skidmore, Owings and Merrill and erected between 1963 and 1965 on a site that faces the main elevation of the great Civic Center (Fig. 90). The loads of the Brunswick are divided between interior shear walls and the exterior frames, from which they are distributed to ten massive columns at the base by means of an immense ring girder that is very likely the largest continuous bearing element ever incorporated into a building. A little over 24 feet deep and 7½ feet thick, it occupies the combined height of the second and third stories and weighs 9,000 tons. Since the thickness of this girder is much greater than that of the framing members above it, and further, since it was necessary to maintain a continuous vertical plane at the inner surface of these members, the outer surfaces of the columns in the truss wall curve sharply inward immediately above the top of the ring girder. In this way an awkward horizontal break in the predominantly vertical pattern is transformed into an easy transition. This feature and the bearing-wall construction make the Brunswick a contemporary counterpart of the celebrated Monadnock Building four blocks to the south of it—two structural and architectural masterpieces separated by seventy-five years of building progress.

The full realization of the economic potentiality of concrete framing came through the development of prestressing. A long and complex history of experimentation in this technique preceded the stage of practical utility. Much of this work was done by European engineers, although the idea was first proposed in 1886 by C. W. F. Doehring of Germany and two years later, independently, by P. H. Jackson of San Francisco. The successful solution to the many problems involved in prestressing came during the decade of the twenties, primarily through the work of the brilliant French engineer and theorist Eugène Freyssinet. This valuable invention entails little change in the form of framing members, but it greatly increases the efficiency of their action. Prestressing is a method of inducing a controlled stress in the member during construction in order to counteract undesirable stresses resulting from the imposition of the working load. In the fundamental case—that of a simple beam under deflection—cables are imbedded in the lower or tension half of the beam and stretched tightly against end plates by means of screws or jacks after the member has set. In this way the tension in the cables is translated into compression in the lower half of the beam. When the member is deflected under load, this precompressive stress cancels the natural tensile stress and thereby places the entire beam under a compression that varies from a maximum at the top surface to a theoretical zero at the bottom. For this technique to be effective, the cable wire must be made of high-quality steel so that it will maintain its tension throughout the life of the structure. The success of prestressing thus depended as much on progress in the metallurgy of steel as in the technology of concrete construction.

Prestressing greatly increases the over-all strength of concrete by fully utilizing its high compressive strength (ranging from 3,500 to 12,000 pounds per square inch) and by reducing the tension to a minimum. As a result, the load may be greatly increased for a given depth of member, or the depth reduced for a given load. Further, prestressing prevents cracking of the hardened concrete, minimizes the creep of the loaded member, and in some cases actually increases its tensile strength. Prestressing is not limited to beams and girders but has been successfully applied to every structural element—columns, slabs, shells, plates, and even the massive blocks of dams. It is frequently combined with the precasting of members, a highly economical method of producing concrete shapes when there is a need for a large number of similar elements. The most recent step toward maximum efficiency is the prestressing of floor and beam elements precast as integral units known as T-beams, in which the upper flange of the beam is widened to act as part of the floor slab.

The economies of prestressing began to appeal to American builders, as we might expect, in the depression of the thirties. The technique was given its first practical demonstration in 1938 in the concrete dome of a clarifier tank designed by Leonard N. Thompson for the water supply system of St. Paul, Minnesota. The tension ring at the base of the dome is reinforced with rods fitted with turnbuckles, by means of which the rods could be drawn tightly around the periphery of the dome and thus be brought to full working stress before being loaded. The compressive stress induced in the concrete ring canceled the tension caused by the horizontal thrust of the dome and minimized cracking by preventing the spread of the concrete under the outward thrust. Subsequent examples of prestressing came in the form of isolated essays until about 1950, when the combination of the post-war building boom and the Korean War taxed the steel mills beyond their capacity and increased the cost of building to the point where drastic economies became a matter of necessity. Within little more than a decade the volume of prestressed construction expanded to an annual rate of a billion dollars. The building industry experienced another phase of adaptive radiation, so to speak, when a new technique was applied to an ever widening range of structures on an ever bolder scale. The chief visual evidence of this progress has been a drastic increase in the span length of conventional framing members. In the Norton Building in Seattle, Washington (1959–60), for example, the main floor girders span 70 feet clear, and at the same time the higher strength afforded by prestressing made it possible to pierce the web of the girders with circular openings for the passage of ducts and conduits.

The most extensive and imaginative use of prestressed framing came with the construction of two large concrete buildings in Honolulu, Hawaii, from 1962 to 1963. The Ilikai Apartments, designed by the architectural office of John Graham and Company and the engineer Howard Lenschen, has been described as the largest prestressed concrete building in the world. The twenty-seven-story structure, in the form of three wings radiating from a central core, was erected entirely of prestressed and precast elements except for the shear walls of the core and the columns of the framed wings. The piles were set in a way that is undoubtedly unique to this building: they were placed in holes drilled through coral ledges to bedrock 120 feet below grade. The structural system of the East-West Center at the University of Hawaii is even more impressive. The complex framing of the building includes continuous spandrel girders of I-section 232 feet long, transverse beams of similar form spanning 128 feet, and I-girders and T-beams to carry the auditorium roof with respective lengths of 124 and 84 feet. A number of architectural and engineering talents collaborated in the creation of this building: I. M. Pei and McAuliffe, Young and Associates were the architects;

Alfred Lee, Donald T. Lo, and K. D. Park were associated structural engineers. The construction of the Center clearly brought the techniques of concrete framing to the scale and flexibility of their most sophisticated counterparts in steel construction.

A combination of prestressing and Vierendeel-truss walls characterizes the construction of the North Carolina Mutual Life Insurance Building at Durham, North Carolina (1964–65), designed by the architect Welton Becket and the engineers Seelye, Stevenson, Value and Knecht. The floors of this fourteen-story building are carried by two groups of seven separate prestressed Vierendeel trusses rather than by continuous frames extending up the full height of the structure, as in the case of the Brunswick Building. The seven pairs of trusses are on opposite sides of the building and extend the 108-foot length of the elevations. Each pair of trusses at any one level supports two floors, and all those on one side are carried in turn by two exterior columns located at the one-third points of the elevation. Since this novel system may be regarded as a set of Vierendeel-truss bridges stacked one on top of the other, it represents a union in a single work of a variety of structural developments derived from both bridge and building construction.

Chapter Nineteen | Concrete Bridges

At the turn of the century the Melan system of reinforcing was in the ascendant for concrete arches, although the more efficient methods of bar reinforcing, introduced by Ransome in 1889, were beginning to gain new attention. For a decade after 1900 the design of arch bridges tended to be conservative, contrary to the development of building construction, which was marked by a bold exploitation of framing in reinforced concrete. By 1910, however, the main line of evolution was moving away from massive construction, with its echoes of the masonry tradition, toward the flattened parabolic curves of narrow ribs, the slender spandrel posts, and the minimal piers that scientific reinforcing was to make possible. Such European innovations as the light slabs of Maillart and the thin-walled box girders of Freyssinet were slow to appear in the United States, and even when they did, they were used conservatively. They lacked the strength and rigidity necessary for railroad loadings, and they often proved too costly for highway purposes. In addition to engineering inertia, a complex of economic and geographical causes lay behind this resistance to the more radical inventions.

In general American practice has called for lower working stresses than those adopted in Europe, and the higher traffic loads have thus required a much greater amount of material to maintain the allowable stress. The higher labor costs in the United States mean that formwork has to be kept as simple as possible, for the erection of forms is largely handwork compared with the mechanized operations of mixing and pouring concrete. Geographical factors include greater wind and snow loads, which reach extremes in the hurricanes along the southeastern coasts and the blizzards of the northern mountains. To these must be added the extreme forces engendered by earthquake shocks in the Pacific coastal region. In the over-all development of structural techniques, however, the economic determinants have always been decisive. The demand for highway bridges, for example, eventually became so great that they had to be erected by methods equivalent to mass production, with the consequence that precast and prestressed girders, deck slabs, and bents ultimately carried the day, while the more costly arch forms fell into relative eclipse.

The conservatism of arch construction at the turn of the century is revealed by the number of large bridges built of plain concrete in forms reminiscent of traditional Roman masonry. Chief among them are the Connecticut Avenue bridge over Rock Creek in Washington, D.C. (1903–4), and the two-mile Long Key Viaduct of the Florida East Coast Railway's Key West Extension (1905–12), rebuilt between 1936 and 1939 by the Public Works Administration to carry U. S. Highway 1. Where reinforcing was adopted, the pattern of development up to 1910 was irregular. The aim of the more imaginative engineers was to create forms in which the concrete and the steel might work in ways most suitable to the properties of each material. The problem was to design reinforcing that would most effectively absorb the tensile and shearing stresses in narrow ribs under moving loads. The Melan system worked well enough, but it was so redundant in the quantity of metal it required as to result in virtually a steel bridge with a concrete cover. The new techniques of bar reinforcing that were incorporated in the Marlborough Hotel and Terminal Station (pp. 241–42) revealed greater promise than the clumsy I-beams of the Melan bridges. Although it is difficult to discover who first returned to the path initially marked by Ransome at Golden Gate Park, the Monier-Wayss and the Kahn systems of reinforcing appear to have led the way. A number of bridges built on the lines of the Illinois Central and Burlington railroads in the first few years of the century incorporated variations on the European technique of arranging reinforcing bars in a two-way grid. The Lake Park Bridge in Milwaukee, Wisconsin, designed and built in 1905 by the Newton Engineering Company, was an early essay involving Kahn's trussed bars. The deck rested on two parabolic ribs spanning 118 feet and reinforced throughout their length with the Kahn bars. The ribs were braced by reinforced beams set between them on the transverse line and by double diagonals in the panels formed by the ribs and the beams.

The end of the experimental phase came in a region which was then far from the major centers of traffic and industry. The engineers of the Oregon State Highway Department, who were to play a leading role in the development of American concrete bridges, made the first of their contributions in the Columbia Highway bridge at Latourelle

Falls, erected between 1913 and 1914 under the direction of K.P. Billner. He was the first American engineer to break entirely with tradition and to treat the concrete arch as a distinct kind of structure comparable in its behavior to the elastic ribs of steel. The bridge consisted of three 80-foot parabolic spans in which the supporting structure was reduced to two narrow ribs, a light spandrel truss in the Pratt form, transverse bracing between ribs, and trussed bents in place of the usual solid piers. The reinforcing in the square-section rib consisted of tension bars bound in a continuous helix that extended the length of the rib, a technique mainly developed by Armand Considère in France. Since the diagonals of the spandrel truss were under tension, they were bonded to the ribs by means of specially designed hooked bars. Although the Latourelle Falls Bridge had to be replaced in 1952 because of increased traffic, it established the essential form of the concrete arch throughout its subsequent history in the United States.

While the Oregon bridge was under construction, its exact antithesis in size, weight, and form was nearing completion in the East. The Tunkhannock Creek Viaduct of the Lackawanna Railroad at Nicholson, Pennsylvania, still the largest concrete bridge in the world, is an enormous Roman arcade that seems at first sight to be thoroughly retrogressive (Fig. 91). Designed and built under the direction of George J. Ray, the company's chief engineer, the viaduct is the most imposing feature of a system-wide modernization that began in the early years of the century and continued until about 1920. The program included new terminals at Hoboken and Buffalo, many new way stations, bridges, and yards, double-tracking, and extensive line relocations undertaken to reduce the grades and curves in the Pennsylvania mountains. The major work in the mountain region was the forty-mile reconstruction of main line known as Summit Cut-off, begun in 1911 and completed four years later. The Summit project quickly attracted attention in the railroad industry for cuts, embankments, and bridges of extraordinary size, the greatest being the huge viaduct at Nicholson. The company undertook this costly program in order to accommodate the rapid expansion of its anthracite tonnage, which was superimposed on a growing volume of miscellaneous freight and an already high density of passenger traffic. Yet this brilliantly engineered transportation artery was used to its full capacity for scarcely two decades. The drastic decline of anthracite mining, the shift of industry from the East to the South and Southwest, depression, and competition drained away much of the Lackawanna's heavy and profitable traffic. The great bridge at Nicholson reminds us of the once flourishing line and the expenditures of money and talent it could then command.

The unparalleled dimensions of the viaduct were established by the need to cross the broad valley of Tunkhannock Creek without loss of elevation from the high grade level of the approach tracks. The over-all length of 2,230 feet is divided into ten semicircular arches of 180-foot span and two flatter abutment arches, buried in fill, of 100-foot span. The track lies 240 feet above the stream and 330 feet above the base of the deepest footing. The suggestion of antiquity in the form of the bridge, carefully enhanced by scoring the concrete to simulate masonry, is a true but superficial impression. The arches, divided into paired ribs that carry heavy transverse spandrel walls to support the deck, contain an elaborate reinforcing that was carefully calculated according to the most advanced scientific standards of the time. The semicircular form of the arch was deliberately chosen on structural grounds in order to minimize the horizontal component of the arch thrust on piers and abutment arches. The process of construction included innovations of permanent value for the erection of concrete bridges. The individual rib was poured on a centering carried by a three-hinged arch truss of steel. When the concrete had set, the whole complex was jacked sideways into position for pouring the parallel member. The bolted truss was then disassembled and re-erected for the next span, the whole process repeated ten times until the structure was completed.

In spite of its old-fashioned appearance, Tunkhannock Creek Viaduct lies in the mainstream of arch development. Highway bridges, with their smaller loads, allowed the engineers to design lighter and more elegant forms; at the same time, the attendant reduction in arch thrust made the flattened parabolic form the general standard. The result was a more organic structure possessing a dynamic quality that is missing in the heavy and static forms of the Lackawanna bridge. The next stage of development came in the area of the Twin Cities, where the Mississippi and Minnesota rivers offered the engineers numerous opportunities to try their ingenuity. The span length of the concrete arch reached a new record in Cappelen Memorial Bridge, built bewteen 1919 and 1923 to carry Franklin Avenue over the Mississippi at Minneapolis. The structure was named after its designer, Frederick W. Cappelen, who was for many years the city's chief engineer and the creator of many of its handsomest bridges. The river crossing of the Cappelen bridge consists of a pair of flattened parabolic ribs with a 400-foot span that carry simple transverse slabs that in turn support the deck. Equally impressive for its sophisticated design is the Mendota-Fort Snelling Bridge over the Minnesota River, erected between 1925 and 1926 after the plans of Walter H. Wheeler. The over-all length of 4,066 feet is divided into twelve spans of paired parabolic ribs. The whole complex of ribs, spandrel posts, and long deck has a finely articulated quality that has seldom been matched in American bridge design.

The longest concrete arch in the United States is the central span of Westinghouse Memorial Bridge at Pittsburgh (Fig. 92). Another masterpiece by V. R. Covell, the bridge was constructed between 1930 and 1931 to carry a major artery over the fantastic chaos of rail lines, streets, and factories that fill the blackened valley of Turtle Creek near the main plant of the Westinghouse company. Five parabolic spans carry the street at a level of 200 feet above the stream. The central pair of ribs span 460 feet between piers, a length that has not been exceeded by an American bridge, although it falls far short of the record now held by the 1,000-foot main span of Gladesville Bridge in Sydney, Australia. The great size of the Westinghouse span and the rigorous quality of its design make it the high point of fixed-arch construction in the United States, but for purity and delicacy of form it stands in second place to the highway bridge constructed between 1931 and 1933 over Bixby Creek near Carmel, California (Fig. 93). The chief engineer of the Bixby span was F. W. Panhorst of the State Department of Highways, and the designing consultant was C. H. Purcell. The bridge represents the culmination of a long

California tradition in concrete design, extending back to the first decade of the century in the pioneer essays of the Los Angeles builders. The tight parabolic curve of the Bixby Creek ribs, set against the sea on one side and a rugged canyon on the other, spans 320 feet and carries the deck 270 feet above the stream. The slender ribs and spandrel bents give the structure a tense and dynamic quality that suggests a modern counterpart of Gothic construction. The only defect of the Bixby Creek bridge is the enlarged depth of the abutment piers, the proportions of which were calculated for an unnecessary visual containment rather than for structural demands.

A European invention of great economic value that has nevertheless remained rare in American practice is Freyssinet's technique of erecting arches by halves under precompression. It was introduced into the United States by Condé B. McCullough, another engineer of the Oregon State Highway Department who became a leading innovator in the design of concrete bridges. He adopted the technique in 1930 for the construction of the Roosevelt Highway bridge over the Rogue River at Gold Beach, Oregon. Each rib of the seven-span structure was erected in halves from the abutment faces. Hydraulic jacks inserted between the halves near the crown before closure precompressed each segment to its maximum stress under normal load and thus prevented the development of secondary stresses arising from the shortening of the rib under traffic loads. In the following year McCullough built the first tied concrete arch to carry Roosevelt Highway over the Wilson River in Oregon (1931–32). Characteristics of the stream banks made the use of conventional abutments hazardous, and the engineer accordingly designed the structure so that the horizontal thrust of the two ribs is translated into tension in the reinforcing of the deck slab, which is suspended below the arches and thus acts as a tie. This could be accomplished only by rigidly uniting the reinforcing in the deck and the ribs.

But McCullough's bridges, like the three-hinged arch, are special forms that have seldom been used in other parts of the country. The fixed concrete arch with abutments continues to be dominant and has been built in considerable numbers for highway spans, although it has been extremely rare among railroad bridges since the pioneer phase of development ended around 1910. Surprisingly enough, however, the form was revived a half-century later for the big bridge of the Western Pacific Railroad over the Feather River near Oroville, California, completed in 1964 as part of a line relocation required by the construction of Oroville Dam. The unusual features of the bridge are the merger of the paired ribs into a single barrel and the presence of full-width vertical slabs in the spandrels to support the deck. The bridge, as a consequence, looks like an expanded and massive version of Maillart's early structures in the Alpine canyons.

With the growth of the highway system and the attendant contraction of the rail network, the concrete girder bridge has now superseded the steel girder as the single most common type of bridge structure in the United States. It exists in a relatively small variety of forms —simple, continuous, and cantilever girders, box girders (usually continuous), and rigid frames—but these have been adapted to an enormous range of dimensions and conditions, with the length of the single span as great as 549 feet. The precasting and prestressing of girders for concrete bridges have brought their construction as close to the methods of mass production as the building arts have yet come. Since the form of any particular type of girder may be repeated hundreds of times, the task of the historian is simply to analyze the basic innovations as they appeared and to point out the conditions under which they function most advantageously.

The concrete girder, like the reinforced concrete arch, began as virtually a steel-beam structure encased in concrete. By 1905, however, the techniques of bar reinforcing were being applied to short-span structures, especially those built for electric interurban railways. Most of these were constructed with simple girders, but cantilever and continuous forms were beginning to appear at the same time. The pioneer work in these closely related forms was probably the three-span bridge built in 1905 by the Marion Street Railway at Marion, Iowa, in which the 50-foot spans consisted of pairs of cantilever girders, each carrying a suspended girder between them. The undersurface of the cantilevers sloped upward from a maximum depth at the pier to a minimum at the end of the suspended girder, a profile that approximated the change in bending moment from the pier to the free end of the cantilever. The 12-inch thick girders of the Marion bridge were reinforced with Kahn trussed bars. A step in the direction of the continuous girder appeared in 1907 in the long viaduct constructed to carry the interurban tracks of the Richmond and Chesapeake Bay Railway over various streets and rail lines in Richmond, Virginia. Reinforcing exactly designed for continuity was used by the engineers of the Burlington Railroad for the bridges built between 1905 and 1910 in connection with the elevation of the company's main line in Chicago. Variations on these types came at intervals during the subsequent history of the solid concrete girder, but again, the main line of evolution lay in the direction of increasing exactitude of form. The continuous girder with an arched lower profile eventually became the dominant type.

Among multispan girder structures the long Western Hills Viaduct in Cincinnati, Ohio, is still one of the most impressive (Fig. 94). Designed by the engineering staff of the city's Department of Public Works, the viaduct was built between 1930 and 1932 to provide a double-level street crossing over the freight yards and the new terminal facilities in the Mill Creek valley. The upper deck, restricted to automobile traffic, was originally 3,500 feet long, but this was extended as a steel structure between 1963 and 1964 to form a connection with the Mill Creek Expressway. The lower level, designed for streetcar loading, extends 2,700 feet. The deck slab and transverse beams of the upper level are carried on a pair of continuous girders, whereas the heavier loads of the lower level are supported by more massive simple forms. Since the completion of Western Hills Viaduct, solid girder bridges in simple and continuous forms have been built by the thousands, the rate of construction accelerating rapidly in the post-war years with the expansion of the urban expressway and Interstate highway systems.

The continuous girder in the form of a hollow box, another achievement of the prolific Freyssinet, was introduced in the United States in 1937 by the Washington State Department of Highways for the bridge over Henderson Bay at Purdy. In this structure the deck slab is the upper cover of the hollow box and thus functions as part of the supporting structure as well as the traffic roadway. The chief structural virtue of the form is that the hollow box possesses greater torsional rigidity than the solid girder while at least equaling the latter in its resistance to bending. A novel variation on the type is the continuously curving box girder that carries Ontario Street over John F. Kennedy Expressway in Chicago. The unusual structure was designed by the engineering firm of Consoer, Townsend and Associates and completed with the opening of the expressway in 1960. The piers supporting the girder constitute a single line of slabs set radially on an alignment that corresponds to the semicircular curve through which the roadway passes.

C. A. P. Turner's system of flat-slab construction was adapted to bridge construction in 1909, with the building of the Lafayette Avenue bridge over the Soo Line tracks in St. Paul, Minnesota. The designer, Thomas Greene, chose the flat-slab structure because restricted overhead clearance at the site made it advantageous to avoid the deep girders of conventional construction. In 1913 the flat-slab bridge was adapted to rail loadings, and ten years later it reached its greatest length with the completion of the Brick Church, New Jersey, viaduct of the Lackawanna Railroad, another famous concrete structure designed and built under the supervision of George J. Ray. Turner's invention has enjoyed a more active life among railroad bridges than any other form of concrete construction. It was most recently employed in the viaduct built to carry the Long Island Railroad over the streets in the commercial center of Hicksville, Long Island (1962–64).

The rigid-frame bridge in concrete anticipated by a few years the comparable form in steel. It was introduced in the United States by Arthur G. Hayden, for many years the designing engineer of the Westchester County, New York, Park Commission. The immediate precedents for Hayden's work were the bridges designed by the Brazilian engineer Emilio Baumgart, the first of which was erected about 1920. The initial arterial system in the county's conservation plan included two scenic parkways, the Bronx River and the Cross County, which were designed to be largely free of grade crossings. Such an ambitious project undertaken by a single county required a great number of bridges and hence the utmost economy of construction. A team of engineers, including Jay Downer as chief, Hayden as designer, and J. Charles Rathbun as consultant, began to investigate rigid-frame structures in 1921 and were soon impressed by their superior efficiency and corresponding saving of material (for the behavior of the rigid frame, see pp. 194–95). On the basis of Hayden's designs, the two parkway commissions constructed seventy-four rigid-frame bridges between 1922 and 1930, the spans ranging from 19 to 99 feet. Hayden developed a standard form for all of them, with only minor variations for skew and curving alignments. The deck slab, a flat plane on the upper surface and arched on the lower, was cast as an integral unit with the downward-tapering abutments. The reinforcing rods were laid longitudinally and transversely near both faces of the slab and abutments because of the complex bending pattern in the rigid frame and the associated changes from tension to compression near the faces of the various members. Bridges set on a skew were also reinforced against torsion—the twisting action resulting from eccentric loads.

The Westchester bridges represent a high point in the scientific design of American concrete bridges; yet even these have been largely superseded by structures composed of separate precast and prestressed girders. Such bridges often seem retrogressive beside the organic forms of rigid frames, but the enormous volume of contemporary bridge construction and the economy of mass-production techniques have offset the saving of material and cost that is possible in the individual rigid-frame structure. The prestressing of girders according to the method perfected by Freyssinet in the twenties was first adopted in the United States between 1949 and 1950 for the construction of the Walnut Lane bridge over Lincoln Drive in Philadelphia (Fig. 95). The designers were the engineering staff of the city's Department of Public Works, with the Belgian engineers M. Fornerot and Gustav Magnel acting as consultants. The deck of the main span rests on thirteen girders, each 160 feet long, which were prestressed by subjecting wire cables imbedded in the concrete to very high tension after the concrete had set. A full-scale girder was tested in the field before construction and failed under a load two and a half times the combined live and dead load on an individual girder in the finished structure.

Although the new techniques of multiple-bridge construction have made it possible to undertake such vast projects as the 41,000-mile Interstate highway system, their potentiality has been most powerfully demonstrated by the building of the single bridge over Lake Ponchartrain between 1953 and 1956. All elements of this 24-mile structure—piles, girders, deck beams, and deck-slab units—were precast and prestressed on the shore, floated to position on barges, and put in place by floating pile drivers and cranes. The result is a perfectly functional and visually uninteresting bridge that is the product of a highly rationalized system of construction—cheap, fast, efficient, thoroughly mechanized, very nearly an industrial process carried on outside the walls of a factory.

Chapter Twenty | Structures of Waterway Control

The control of stream flow from end to end of a waterway system for a multiplicity of interrelated purposes is perhaps the most valuable technical accomplishment of the century, and it is to a considerable extent the achievement of American engineers since World War I. The various stages in the evolution of this technological complex—water storage on a large scale, maintenance of adequate flow for navigation, irrigation, and hydroelectric generation—were introduced before 1900, but they could not be brought together as a functioning unit until two major engineering concepts could be fully implemented in practical terms. One is the complex of special dams designed and distributed in such a way as to control stream flow over an entire drainage system, and the other is the multipurpose dam, which combines in one structure the regulation of flow for navigation, irrigation, power generation, and flood control. These aims could be realized on a regional scale only through the entry of the federal government into the activity of dam construction. Most of the big dams are thus products of the social revolution that led to the active role of the government in the economic life of the nation.

Stream control for the maintenance of navigation antedates the more elaborate multipurpose programs first undertaken by the Bureau of Reclamation. Systems of navigation dams are concentrated in the Midwest, the agricultural and industrial heartland where the main inland waterways are located. The construction of these dams has always been the responsibility of the Army Corps of Engineers, who have as their goal the transformation of unruly rivers into dependable canals. The first streams to be brought under control were the naturally navigable arteries, the Ohio, Illinois, Mississippi, and Missouri rivers. The earlier construction of the Chicago Sanitary and Ship Canal (1891–1900) connected the vast Mississippi system with the Great Lakes, and the opening of the St. Lawrence Seaway in 1959 joined these to the Atlantic Ocean. The establishment of the Tennessee Valley Authority and the extension of the Army Engineers' activities added the Allegheny, Monongahela, Tennessee, Cumberland, Warrior, Columbia, Snake, and Arkansas rivers, the last of which reached the planning stage in 1964, with construction beginning in 1966. The ultimate results have been the creation of the greatest system of inland waterways in the world and the attendant resurgence of waterway transportation, which had been nearly killed by the railroads around the time of the Civil War.

The first major step in the implementation of the Army Engineers' program was the canalization of the Ohio River, initiated in 1879 and completed in 1929. The successful operation of this plan required an associated system of flood-control works in the major tributaries of the Ohio, which was undertaken shortly after the disastrous floods of 1913 and is still in the process of completion. The original group of forty-six dams in the Ohio River is composed for the most part of a type known as the wicket or shutter dam, which is restricted to the Ohio and Illinois rivers. The wicket dam is in essence a movable weir of steel-bound timber plates hinged at the base and supported on the downstream face by a hinged steel bracket. The hinges are fixed to a concrete slab laid on the stream bed. The lock for the passage of vessels is located at one of the shore ends of the dam. This simple low-head installation normally acts to impound a navigation pool, but in times of high water the hinged panels can be laid flat on the bed, allowing vessels to pass directly over the dam. The chief defect of the wicket dam is its rather flimsy construction, which makes for a short life and a maintenance problem, although repair is usually a simple matter of replacing a broken plate by means of a floating derrick.

Included among the Ohio River structures are two dams of more advanced design and more substantial character known as rolling crest dams. The one constructed at Gallipolis, Ohio, between 1936 and 1937 is a typical example (Fig. 96). The fixed structure is a concrete wall extending across the stream and supporting a series of concrete piers. Between each pair of piers is a hollow steel cylinder that can be raised to admit the passage of water under the cylinder or lowered to act as a conventional impounding gate. The chief virtues of the rolling crest dam are the ease with which the geared roller can be moved, the exact mechanical control of the movements and hence of the volume of water admitted through the structure, and the relatively large width of unobstructed opening provided by the rigid cylinder. The origin of this ingenious form of water control is obscure.

It was developed in the early years of the century as an alternative to the horizontally hinged gate and was adopted by the Bureau of Reclamation in 1913 for the first dams of the Colorado River Project. The type is best suited to broad, relatively shallow streams carrying a large volume of sediment. The Army Corps of Engineers used it exclusively for the canalization of the Mississippi River above Alton, Illinois (1930–48) and for several installations in the Illinois Waterway (1930–39), and they have adopted it, along with the conventional gated spillway, for the replacement of the wicket dams in the Ohio River. This last program, initially planned in 1958 and now in process of implementation, will eventually replace the forty-four wicket dams with seventeen of the roller and gated types, making a new total of nineteen for the entire river.

Dams designed for the maintenance of navigation are usually low-head structures in which the movable parts are the visually prominent features. At the other end of the scale are the high-head storage dams built for flood control, irrigation, or power generation or for all three together. Among storage dams, the largest and most complex have been built by the Bureau of Reclamation, established by the Congress in 1902 as an agency of the Department of the Interior. The establishment of the bureau proved to be one of the most far-reaching and revolutionary acts ever accomplished by an American Congress. Its original purpose was the construction of dams, canals, and appurtenant structures to provide irrigation for the arid regions of the plains and the intermountain deserts, but the subsequent expansion of its activities into power generation and flood control raised it to the level of a major factor in the American economy, one of such importance as to make it the very basis of civilized existence in many western states.

Between the midwestern prairies and the coastal mountain ranges lies an enormous and diversified region all parts of which have one feature in common, namely, inadequate rainfall for the maintenance of agriculture. On the treeless plains an annual precipitation of eighteen inches is barely enough for subsistence farming; in the deserts west of the Rockies there are areas so dry that not even the hardiest desert vegetation can survive. To the pioneers heading for the Pacific coast it was a wasteland offering nothing but terrors. The geologists who first explored it systematically, however, discovered that great areas of the region are covered with volcanic soil rich in the minerals necessary to support intensive agriculture. Theodore Roosevelt, who proposed the reclamation program, was the first president to see the unlimited promise of the region and the obstacles that stood in the way of its realization. Before 1905, however, concrete and hydroelectric technology had not advanced to the stage required for the construction of storage and power dams in the western valleys, and it was only on this technical basis that the engineers of the Bureau of Reclamation could initiate the program that was to produce the largest structures ever made by man.

The first reclamation dam to embody many of the features of the western installations is Roosevelt Dam, built in the Salt River between 1906 and 1911 near Phoenix, Arizona. It was the first to outdistance all its predecessors in size, but it was also the last in the tradition of masonry construction. Roosevelt is an arch-gravity dam in form—that is, one that sustains the load of impounded water both by weight and by the compressive action of the arch—and it is constructed in the traditional way, with a concrete and rubble core covered by rough-faced blocks that are laid in a mortar so coarse as to be a concrete. Its height of 284 feet and crest length of 1,125 feet established a record at the time. Although its masonry structure belongs to the past, its subsidiary facilities and the process of construction include many features that were to appear with variations in all subsequent installations of the bureau. The mass of the dam was built up from bedrock, which was first exposed by diverting the river through a tunnel cut in the canyon wall and then maintained in the dry state by laying down cofferdams above and below the dam site. The electric power generated at the dam is mostly used to operate the pumps that move the water into the diversion system, but the excess power is sold to the city of Phoenix according to a provision in the authorizing act that first put the Bureau of Reclamation into the business of producing and selling electricity.

The bureau had embarked on the construction of concrete dams in 1905, but it was the Arrowrock installation, built between 1912 and 1916 in the Boise River near Boise, Idaho, that first demonstrated the possibilities of the concrete-arch form for waterway control. The exclusive purpose of Arrowrock Dam was to provide irrigation for 166,000 acres in the volcanic desert of southern Idaho. The original height of the dam, 384.5 feet, established a world record at the time of its completion, and it was raised another 5 feet between 1936 and 1937 to meet an increased demand for irrigation water. The arch-gravity form of Arrowrock became the standard for dams located in deep narrow canyons in which the rock of the side walls is sufficiently dense and homogeneous to sustain without leakage the immense lateral thrust of the arched mass of concrete. Since such dams are also gravity structures, they act as vertical cantilevers as well as arches in lateral compression.

The rapid expansion of the Bureau of Reclamation's activities and the multiplication of its works in the highly diversified topography and rock formations of the western streams required the use of special forms of dams adapted to peculiar hydraulic and geological conditions. Certain of the types are radically different from the conventional monolith of concrete and represent novel forms derived from vaulted and buttressed structures. The first of these is the multiple-buttress dam, invented by N. F. Ambursen in 1903 but not adopted by the bureau until the construction from 1926 to 1928 of Stony Gorge Dam in Stony Creek near Orland, California (Fig. 97). In this type the water load is supported by a row of concrete buttresses in the shape of triangular slabs set with their narrow edges facing up and downstream. These carry reinforced concrete slabs spanning between the upstream faces of the buttresses and inclined in the direction of the stream flow. The succession of slabs, acting as broad beams, impounds the water and transmits the load chiefly as a compressive force to the buttresses. The multiple-buttress dam is a highly articulated form that requires a minimum of material and offers the great advantage that any failure of the structure will be limited to a single bay because of the construction in separate load-bearing units. The bureau adopted the form for Stony Gorge because of the presence of a fault line at the south abutment of the dam, where the complex strata of shales and sandstones had been displaced eastward about 150 feet. Since the region is subject to earthquakes, the possibility always remains of further movement along the fault. The dam is moderate in size—142 feet high and 868 feet long—but what it lacks in mass is offset by its elegant articulation, which seems precariously light for the depth of water behind it.

The strangest looking of all the structures of waterway control is the multiple-dome dam, which was first employed by the bureau for the Coolidge installation, constructed between 1926 and 1928 in the Gila River near Globe, Arizona (Fig. 98). In this form, which was invented by the bureau's engineers, a series of narrow buttresses support semidomes of ellipsoidal shape. The maximum transverse span of the dome (the minor axis of the ellipsoid) is at the base of the dam, and the convex curvature faces upstream. The form possesses two virtues inherent in the shape itself: the two-way compressive stress of the doubly curved surfaces opposes both lateral and downstream bending, and the great rigidity of the connected domes increases stability against overturning. Like the multiple-buttress, the multiple-dome dam is limited to structures of intermediate size. The over-all height of Coolidge Dam is 249 feet, for a straight-line length at the top of about 600 feet.

One of the most impressive of American dams for its graceful and organic form is Bartlett Dam, built by the bureau between 1936 and 1939 in the Verde River near Phoenix, Arizona (Fig. 99). The design of this particular installation was based on theories previously developed by the Norwegian engineer Fredrik Vogt, although the multiple-arch had been introduced into American practice by John S. Eastwood in 1909. The fundamental idea is again the principle of separate buttresses, but in this case they constitute the supports for a succession of reinforced concrete vaults whose axes are inclined in the direction of stream flow, the convex curvature facing upstream. The water load is transmitted to the buttresses by compression in the vaults, and the lateral horizontal thrust of each vault is balanced by that of its immediate neighbor, except at the ends, where the thrusts are sustained by the abutments in the canyon walls, which must be dense enough to carry the great compressive force. The 273-foot height of Bartlett was long thought to be standard for the multiple-arch type, but this limitation was swept away with the design of the spectacular Manicouagan Five Dam of Hydro-Québec (1963——), the 703-foot height of which makes it one of the highest dams in North America.

For the greatest installations of the bureau the more conventional gravity and arch-gravity dams have been the rule. Among the latter, the largest and most famous is Hoover Dam, located in the Black Canyon of the Colorado River near Las Vegas, Nevada (Fig. 100). Hoover was the first dam specifically designed to generate power in the quantity necessary for modern urban needs, with the consequence that all the essential features of the design, construction and operation of the hydroelectric-irrigation-flood control project are here embodied on a tremendous scale. Plans to build a major storage dam in the Colorado River were first considered by the bureau staff in 1904, two years after the agency was established. The engineers carried on hydrological investigations and surveys of potential sites for twenty-four years before deciding on Black Canyon as the best location. The act authorizing the construction of the dam was signed by President Coolidge in December, 1928. The research staff of the bureau undertook an extensive program of load and hydraulic tests on models of the dam and spillway for the two-year period 1929–31. The contract for construction was let in the latter year, and the vast structure was completed in 1936, the highest dam in the world at the time and the largest generator of electric power.

In many respects the Black Canyon site is ideal for the construction of a high-head power and storage dam, but the scale of the operations and the unruly river posed unprecedented difficulties. The narrow canyon, its walls rising almost vertically for the first 500 feet, and the hard, dense, impermeable volcanic rock made it possible to raise the dam to its great height of 727 feet, but the task of pouring 3,250,000 cubic yards of concrete and squeezing all the appurtenant structures into the narrow space reached staggering proportions. Since the site was inaccessible, before construction could begin the contractors had to build a railroad line, heavy-duty roads, and an electrical transmission line from Las Vegas to the canyon rim. The violent flash floods of the Colorado required cofferdams of record height, and the stream had to be diverted around the site by drilling and blasting tunnels through both walls of the canyon. The heat generated by solar radiation reflected from the smooth walls became so intense that operations frequently had to be suspended because workmen could not handle machines and tools. The river course in the canyon is so constricted that many of the subsidiary facilities had to be built outside the body of the dam: the intake works were placed in separate towers above the upstream face —four skyscrapers of braced concrete buttresses; the penstock tubes and the spillway discharge tunnels were drilled through the side walls; and the huge powerhouse had to be squeezed into a tight U-shaped plan at the downstream base of the dam. The whole process of design and construction was a work of sheer technical virtuosity, an exhibition of scientific engineering on the highest level. The operation of Hoover Dam has become essential to the life of Los Angeles. The storage reservoir holds much of the water distributed by the Metropolitan Water District, and the powerhouse generates about 4,000,000,000 kilowatt-hours of electricity per year to meet part of the city's demand.

Before the Hoover project was completed the Bureau of Reclamation began the construction of Grand Coulee Dam, a straight gravity structure located in the Columbia River near Spokane, Washington (Fig. 101). Every feature of this grand creation of man's building prowess reaches the superlative degree: it is the largest concrete dam and includes a spillway of the greatest capacity; it is the largest structure built from man-made materials and the first after nearly forty-six centuries to exceed the Great Pyramid of Khufu in volume. The initial survey for the Columbia Basin Irrigation Project was made in 1921, but construction was not begun until 1933, when President Franklin Roosevelt's secretary of the interior, Harold L. Ickes, began to implement his ambitious program of public power for the Pacific Northwest. The dam was closed and the first group of generators placed on the line in 1943, but nearly another decade passed before the power installation was completed. The dam took its name from an ancient lava-walled valley that was eroded down to its granite bed by the Columbia River when the stream was temporarily diverted by a glacier. Subsequent erosion in the previous course of the river left the dry valley floor high above the present water level in the stream. This natural formation, known as the Grand Coulee, is ideally shaped for an equalizing reservoir, requiring only the construction of control dams at its narrow ends.

The entire Grand Coulee project embraces an extraordinary range of subsidiary elements. In addition to the dam proper, there are two powerhouses flanking the central spillway, a wing dam above the upstream face to protect the pumping installation, two pumping plants, two earth and rock-fill dams in the equalizing reservoir, irrigation control works, and several thousand miles of pipes, tunnels, canals, and transmission lines. The gigantic dam stands 553 feet above the foundation, extends 4,173 feet along the top, and contains 10,585,000 cubic yards of concrete. The overflow spillway was designed to discharge 1,000,000 cubic feet of water per second, the normal flow of the Mississippi River at New Orleans. The spillway is one of the few that are regularly used, since the Columbia floods its valley every June as the result of melting snow in the Canadian Rockies. In early summer a major part of the flooded river slides easily over the parabolic crest of the spillway and drops 350 feet to the upturned surface of the toe, the volume of water frequently greater than that of the Niagara River and the descent twice the height of the Falls. For the operation of the whole system, the water of the Columbia plays a double role: the smaller volume passes through the penstocks in the body of the dam to the eighteen turbines of the powerhouses; the larger volume flows into the pumping plant, where a battery of the largest pumps ever built lifts it 295 feet into the reservoir in the Coulee. Only in time of flood is any of the precious water wasted. Otherwise it moves silently through hidden passages, ceaselessly rotating the turbines or flowing up the face of the lava cliff to pass eventually to the wheat farms of the Big Bend region. The human scale is lost before the bare lava rock of the area; yet nowhere is there a greater example of creative technology employed to serve fundamental human needs.

If many of the dams that followed Grand Coulee seem anticlimactic in size, they reveal steady progress in engineering design and methods of construction. The most sophisticated in their form are two installations that constitute major units in the Colorado River Storage Project. One of them, Glen Canyon Dam (1956–66), is located in the torturous canyon of the Colorado that extends across the Arizona-Utah boundary (Fig. 102). This structure is one of the few American dams that sustain their load by arch action alone, and it is the highest of that kind so far built in the United States, standing 710 feet above the foundation. The difference between Glen Canyon and an arch-gravity dam like Hoover is revealed by a single dimension: in spite of the fact that the newer dam is 55 per cent greater in volume than the older because of the greater width of the canyon, its base thickness of 340 feet is little more than half that of its predecessor. Another dam in the Colorado River Storage Project that differs even more radically from conventional forms is Morrow Point Dam (1963–68) in the Gunnison River near Gunnison, Colorado, the first thin-arch double-curvature dam in the United States. The elaborate designation, however awkward, exactly describes the form and mode of action. The envelope of concrete is a trapezoidal segment of an enormous dome that gains its strength and rigidity from the vertical and horizontal curvature, which in turn develops a two-way compressive stress. Again, the major dimensions reveal the drastic change in proportions from those of standard forms: although Morrow Point stands 469 feet high, it is only 52 feet thick at the base. Translated into ratios, the differences are astounding: Morrow Point is nearly two-thirds the height of Hoover, but its volume is slightly more than 10 per cent and its base thickness is only 8 per cent of the equivalent dimensions of the older dam. The thin-arch dam is a European invention, and its belated importation into the United States shows once more the tendency of American engineers to follow the leadership of their more experimental and more scientifically minded fellows in Europe. As we have repeatedly noted, this dependence has been a constant characteristic of American building from its colonial origins to the present time.

The most extensive system of waterway control in which the Bureau of Reclamation is involved is the Pick-Sloan Plan for the Missouri River, a joint activity of the bureau and the Army Corps of Engineers inaugurated in 1944. By 1964 this program included twenty-seven dams in the Missouri basin, all of them full multipurpose installations. Among them are the four largest dams ever constructed, earth-fill barriers with concrete powerhouses and spillways that stand in the broad valley of the main river from Fort Peck, Montana, to Pierre, South Dakota. The smallest, Fort Randall Dam (closed in 1956), is 53,000,000 cubic yards in volume, and the largest, Fort Peck (closed in 1940), extends 21,000 feet along the crest and contains 125,628,000 cubic yards of earth. With modern earth-moving machinery, man can now rival the geological forces that shape the surface of his planet.

The third of the dam-building agencies of the federal government is the Tennessee Valley Authority, a unique institution established in 1933 to achieve total waterway control, conservation of soil and forests, and the orderly development of resources in the region of the Tennessee River. In these respects the authority has been an astounding success, its achievement such a brilliant exhibition of political, technical, and scientific skills that the Congress has never dared repeat the masterpiece of legislation that established it. The designers of the physical plant of TVA have drawn heavily on the experience of the Bureau of Reclamation, but their work differs from the western installations in two respects: first, the controlling devices constitute an interlocking system organic to the natural processes of the valley itself, and second, the architectural character of the structures and the enhancement of the surrounding natural beauties (the concept known as total design) represent the highest level of design in any technical-industrial complex of the United States.

The background to the concept of TVA lies less in engineering history than in the development of the conservation movement in the United States. The beginning on a governmental level is the federal conservation program submitted by Gifford Pinchot to President Roosevelt in 1907. At first concerned mainly with water resources, Pinchot expanded the program in the succeeding two years to include timber, soil, and mineral resources. His central idea was the concept of the river basin as an organic unit in which the waterways, the regional physiography, the soil, vegetation, and fauna followed their respective natural processes in an ecological equilibrium. The proper exploitation of these resources, therefore, was to be carried out by means of a balanced multipurpose program embracing these major features: stream control for flood prevention, navigation, power, and recreation; replanting and selective cutting of forests; the preservation of meadowland, the restoration of soil fertility, and the prevention of future soil exhaustion. This profound and far-sighted plan for man's responsible partnership with nature was embodied in the recommendations of the National Conservation Commission published in 1909, but it was not implemented systematically on a regional basis until the establishment of TVA in 1933.

The technical features of the authority's program include two basic parts, namely, the control of the main river for flood control and power as well as navigation, and the storage of water in the mountain tributaries for the retention of flood water and the generation of power. Proposals for the maintenance of year-round navigation in the Tennessee River go back to 1824, when John C. Calhoun first recommended a program of dredging and channel control. In the following century the Army Corps of Engineers carried on a hydrological survey of the river that provided the necessary data for a plan of canalization. The appropriate controlling structure was a low-head concrete gravity dam of sufficient height and weight not only to maintain a navigation channel but to impound water for flood control and power generation. The first positive step was the construction by the Army Engineers of Wilson Dam at Muscle Shoals, Alabama, between 1918 and 1925. The prototype of this structure, now the largest producer of hydroelectric power in the eastern United States, is Keokuk Dam in the Mississippi River at Keokuk, Iowa (1910–13). The construction of Wilson led to a long and bitter controversy over the disposition of the power facilities, but finally, the dam was saved to be incorporated in the TVA system largely through the efforts of Senator George W. Norris. The remainder of the system in the Tennessee River was completed through the construction and purchase of facilities by the authority itself, which began its own program with Wheeler Dam, constructed from 1933 to 1936 near Decatur, Alabama (Fig. 103). Typical of the main river structures, Wheeler is composed of a long, low base of trapezoidal section carrying a row of slab piers supporting the roadway at the top. The outdoor powerhouse (original with TVA) and the lock are at the ends of the dam, and the gate-controlled overflow spillway flanks the powerhouse. The last and longest of the main-river series is Kentucky Dam, built near the mouth of the Tennessee between 1938 and 1944.

The high-head tributary dams differ widely in size and in details, but all are straight gravity dams of concrete or earth and rock-fill. The first of these is Norris Dam, constructed between 1933 and 1936 in the Clinch River near Knoxville, Tennessee, and the last is Melton Hill Dam, located immediately below Norris in the same stream and closed in 1963. The most impressive of the authority's structures are located in the tributaries that rise on the western slopes of the Great Smoky Mountains. Among these, Fontana Dam (1942–45) is the largest and handsomest (Fig. 104). Its highly refined engineering is superbly expressed in the simple monolith of the dam and the glass-walled powerhouse at its base. The entire hydroelectric system of TVA embraces thirty-two dams, of which twenty-one were built by the authority. Associated with these are seven dams built by the Army Engineers in the Cumberland River and its tributaries, of which the most recent is Barkley Dam (1958–65) near Paducah, Kentucky. By 1940 the demand for electric power in the Tennessee basin was beginning to outstrip the hydropotential, and the authority accordingly embarked on a program of constructing ten steam-generating plants, which in 1966 supplied 83 per cent of the total power generation of 81.1 billion kilowatt-hours. Several of these installations are the largest steam plants in the world. The one at Bull Run, Tennessee, for example, houses a single turbine-generator unit with a capacity of 900,000 kilowatts, greater than the entire generating capacity of the Tennessee basin when the authority was established in 1933.

103. Wheeler Dam, Tennessee River, near Decatur, Ala., 1933–36. Tennessee Valley Authority, engineers. Wheeler is typical of the low-head, main-river installations of TVA.

104. Fontana Dam, Little Tennessee River, near Bryson City, N. C., 1942–45. Tennessee Valley Authority, engineers. In its sophisticated design and its harmonious relationship with its mountain setting, Fontana Dam best exemplifies the technical and social principles of TVA.

105. Domed filter covers, sewage treatment plant, Hibbing, Minn., 1939. J. C. Taylor, architect; C. Foster and the Roberts and Schaefer Company, engineers. An early example of shell domes in the United States.

106. Lambert Field Terminal, St. Louis, Mo., 1954–55. Hellmuth, Yamasaki and Leinweber, architects; William C. E. Becker, engineer. The terminal building at Lambert Field is a large-scale example of ribbed shells in the form of groined vaults.

105

106

107. Assembly Hall, University of Illinois, Urbana, Ill., 1961–62. Harrison and Abramovitz, architects and engineers. Cross section. Folded-plate and shell-dome construction are combined in this big university assembly hall.

108. Research Building, S. C. Johnson and Company, Racine, Wis., 1947–50. Frank Lloyd Wright, architect; Wesley W. Peters, engineer. Section through the tower. The research tower of the Johnson company is the first building in the United States to embody pure core-and-cantilever construction.

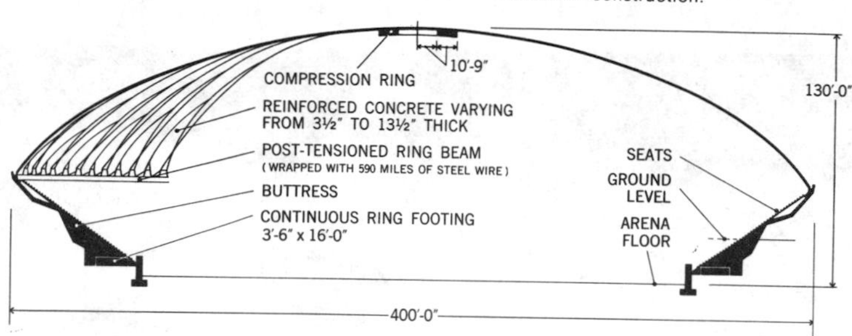

107

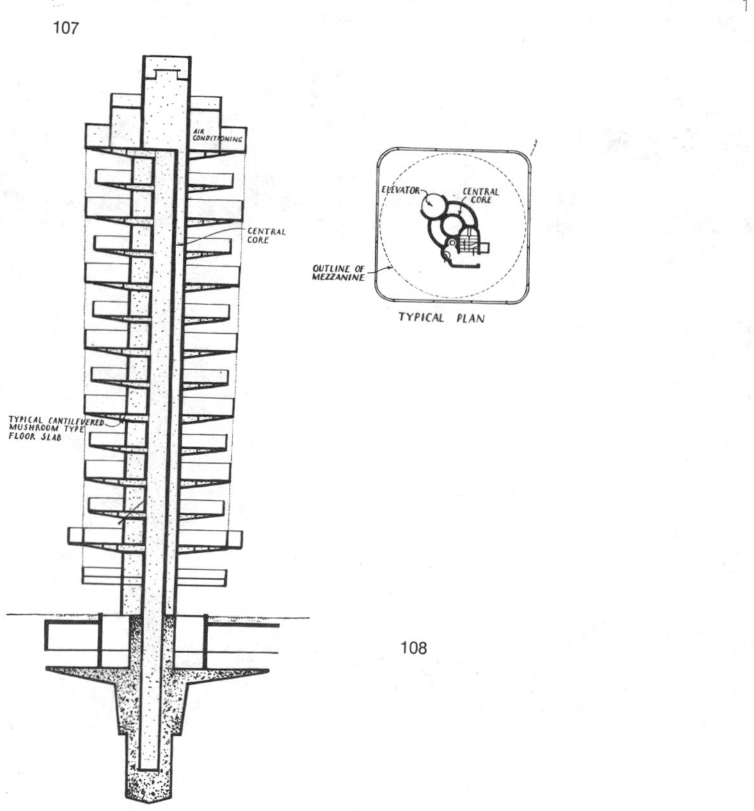

108

109. Solomon R. Guggenheim Museum, New York City, 1956–59. Frank Lloyd Wright, architect; Wesley W. Peters, engineer. Interior view of the exhibition rotunda. The spiral ramp of the museum appears to be self-supporting but is actually carried by a hidden system of ring beams and slab-like piers.

110. Marina City, Chicago, Ill., 1960–64, 1965–67. Bertrand Goldberg Associates, architects and engineers. All the techniques of concrete construction are embodied in Marina City, but the most conspicuous are the cylindrical cores and ring colonnades that carry the flower-shaped floors of the apartment towers.

111. Convention Hall, Democratic Party, Houston, Tex., 1928. Franzheim and Dowdy, architects; Klingenberg and Kelly, engineers. The lamella roof during construction.

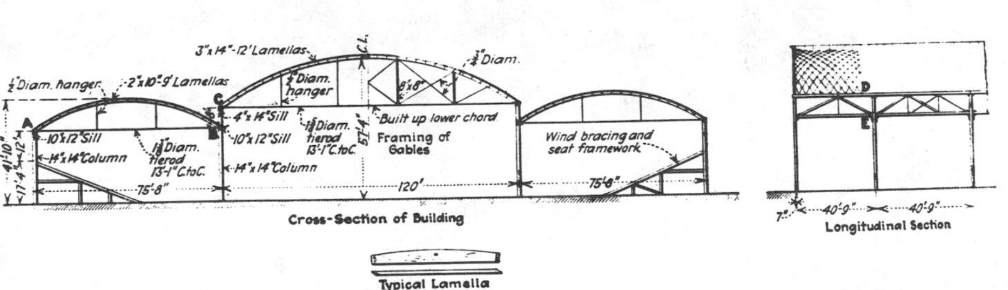

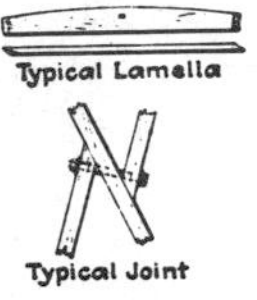

112. Thomas Jefferson High School, Port Arthur, Tex., 1962–63. Caudill, Rowlett and Scott, J. Earle Neff, architects. Timber arches of the roof during construction.

Chapter Twenty-one | Concrete Shells and Related Forms

Except for dams, the concrete structures we have so far considered fall into two broad categories: the larger consists of rectilinear systems of frames or slabs, or combinations of the two, and the smaller includes a variety of arched forms such as ribs, barrels, domes, and vaults. All these structures, whether rectilinear or curved, are subject to varying degrees of bending, so that the individual members must always be heavily reinforced and relatively massive in construction. Shortly after the turn of the century, however, a few European engineers realized that the efficiency of concrete could be greatly increased if the shape of a structural element under light load were such that its curvature alone would provide the necessary rigidity and strength. The most commonplace analogy may be found in the behavior of a sheet of paper. Flattened out and held at one end, it will bend downward under its own weight; curved into the longitudinal segment of a cylinder, however, it will sustain itself like a little cantilever. This ideal of two-dimensionality may be attained if loads can be translated into forces acting wholly within the envelope of the curved form and never at right angles to its tangent planes. The theory of the perfect arch expresses the same idea: if the curve is a catenary, and if the load is statically and uniformly distributed along the curve, the arch will stand if a sufficiently rigid material is concentrated along its axis. It is this reasoning that lies behind the revolutionary technique of thin-shell construction.

The background of shell forms is extremely complex and extends far into the past. Shell-like structures in masonry may be found in the brick semidomes and abutment vaults of Hagia Sophia Cathedral in Istanbul (A. D. 532–37) and, in more advanced form, in the flattened brick vaults erected by the Catalan builders of the Middle Ages. A similar structure is the vault webbing of the Gothic cathedral, although exactly how the load was distributed between the webbing and the ribs is still a matter of controversy. But these exhibitions of the mason's virtuosity had no continuous history, and it is doubtful whether modern builders in concrete paid any attention to them. The preparation for building in shell forms came chiefly with the construction of domed, vaulted, and other curved elements in European and American building during the early years of the twentieth century.

In the prolific period 1900–1915 the enthusiasm for concrete construction induced the engineers and architects of the United States to try their hands at a great variety of ribbed, domed, and vaulted forms. The Marlborough Hotel (1905–6) and the Traymore Hotel (1914–15), both in Atlantic City, include a number of domed enclosures within their structural frames. These were hemispherical envelopes of concrete resting on reinforced ribs. More advanced was the system of concrete vaults and ribs that supported the dyehouse floor in the Botany Worsted Mills at Passaic, New Jersey (*ca.* 1906); the ribs lay on the diagonal lines of the bays, forming a row of X's in plan, like the diagonal ribs of a Gothic nave vault. Concrete groin vaults without ribs were introduced in 1909 into the design of water filtration plants, of which the most striking example is the cover of the Baldwin Reservoir in Cleveland, Ohio (1922–23). Most fully emancipated from masonry precedents are the cylindrical grain elevators that have been built in concrete since the end of the nineteenth century. A typical early example is the Great Northern Railway's elevator at Superior, Wisconsin (1909), the individual container of which stands 110 feet high with a wall thickness of only seven inches. All these works held valuable lessons for the full exploitation of the plasticity and strength of curved concrete forms, but in the United States they went largely unheeded.

The difference in thickness between a shell and a conventional domed or vaulted form is somewhat arbitrary, but the difference in structural action may be marked. The shell is regarded as a surface structure or two-dimensional form, that is, one that in theory has no working depth. When the ratio of thickness to span for a curved envelope approaches a limit of 1:500, or 2.4 inches for 100 feet, the envelope is regarded as a shell, and in a few cases the ratio has been even smaller. The French and German engineers who were primarily responsible for the development of shells seem to have arrived at the technique more by intuition than by progressive experimenting with the dimensions of existing concrete forms. The first shell vault was built in 1910 for a railway station in the suburb of Bercy in Paris, and in 1916 Freyssinet gave a powerful demonstration of the form in the celebrated airplane hangar at Orly, near Paris. During the twenties German engineers began to construct domes and multilobe vaults as thin shells. Before another decade passed, builders in Europe and Latin America had expanded the vocabulary of shell forms to a variety of novel shapes derived from the geometry of conic sections. When shell construction finally came to the United States during the depression of the thirties, it did so strictly as a European importation.

The first American essay was an experimental work designed for an exposition, motivated partly by the aim to produce a novelty and partly by the need to build as economically as possible. The pavilion of the Brook Hill Farm Dairy Company at the Century of Progress Exposition in Chicago was a small rectangular enclosure roofed by a multispan shell vault. Constructed for the second season of the fair (1934), the building was designed by the architect Robert Philip and the engineering firm of Roberts and Schaefer Company. The roof consisted of five contiguous elliptical vaults with their axes running in the direction of the short dimension of the rectangle—a multilobed vault, as it is usually called. Each vault, with a thickness of three inches, spanned 14 feet, for a total clear span of 70 feet. The vaults were held rigid by an edge rib curved to the shape of the elliptical section and by light longitudinal beams on the lines where the adjacent vaults met.

The number of shell structures and the variety of forms increased slowly at first. The shell dome in the traditional hemispherical form came with the construction of Hayden Planetarium in New York City (1934–35), the work of the architects Trowbridge and Livingston and the engineers Weiskopf and Pickworth. The metal shell which constitutes the projection screen for the planetarium instrument is fixed to the inner surface of a concrete shell with a thickness of 3 inches for a diameter of 80 feet 6 inches. Three years after the completion of the planetarium the first single-span shell vault was constructed to roof the building of the Philadelphia Skating Club and Humane Society. Roberts and Schaefer were again the engineers, and E. N. Edwards was the architect. The clear span of this vault is 105 feet and the minimum thickness (at the crown) 2⅝ inches; however, a series of transverse ribs play a considerable role in maintaining the rigidity of the structure.

During the great expansion of concrete construction following World War II, shells came to play a major role in single-story structures where intermediate supports are impracticable. Once more it was the economy of the technique that proved to be the decisive factor in its wide adoption. The flattened ellipsoidal dome had appeared in 1939, when the town of Hibbing, Minnesota, built a group of domed filter covers for its sewage treatment plant (Fig. 105). The engineers of the Roberts and Schaefer Company, working in collaboration with the architect J. C. Taylor, were the designers. The dimensions of the covers reveal the steady progress in the refinement of form: the thickness is 3½ inches for a diameter of 150 feet.

Groin vaults obtained from the intersection of elliptical cylinders characterize the greatest American work of shell construction for over-all size and sweeping interior space. The terminal building of Lambert Field in St. Louis (1954–55) has a length of 415 feet without an intermediate support, and the diagonal-rib span of each of the three vaults is 164 feet (Fig. 106). William C. E. Becker was the engineering designer of this masterpiece, and Hellmuth, Yamasaki and Leinweber were the architects. The great volume of building accompanying the St. Louis redevelopment program included the first hyperboloidal shell, the odd spool-like cover of the Municipal Planetarium in Forest Park, completed in 1963 from the plans of Hellmuth, Obata and Kassabaum. The same year saw the completion of the largest example of the doubly-curved surface known as the hyperbolic paraboloid (or hypar), the roof of the Edens Theater in Northbrook, Illinois, designed by Perkins and Will. As remarkable as its great size is the fact that it is supported at only two points, the ends of its transverse or shorter axis. The saddle-shaped hypar is the most rigid of shell forms because it has a curved section in each of the three perpendicular planes. In spite of this complexity, it is a ruled surface, that is, one made up of an envelope of straight lines, a feature that offers the great advantage of allowing the use of straight boards for the formwork in which the concrete is poured.

Closely related to shells in the thickness-span ratio and in the mode of action are folded plates, which would be more accurately described as pleated rather than folded. Again the sheet of paper provides a good analogy: if the paper is folded back and forth along a number of parallel lines, the resulting pattern of ridges and valleys has a much higher rigidity than the flattened sheet. The same effect may be achieved by bending the paper into a series of parallel waves, a technique adopted for the manufacture of corrugated paper containers. The reason for the increased rigidity is that the sloping or curving surfaces provide an increased depth of working material in the line of the load and thus act like a system of little parallel ribs. For this reason folded and corrugated shapes, unlike shells, can sustain a moderately high bending force. It seems clear that folded plates were anticipated in the skylight structure of Terminal Station in Atlanta (see p. 242), but there is no evidence that the novel details of this building had any influence on American practice. Once more it was the European engineers who led the way, with Freyssinet in the vanguard by virtue of the corrugated-shell construction of the Orly hangar.

Folded plates as covers for an entire enclosure did not appear in the United States until about 1950. The form offers special advantages for airplane hangars, where the primary requirements are extensive clear space and maximum overhead clearance within the enclosure. Since there can be no peripheral supports in hangars to interfere with the movement of planes into the roofed area, the cover is best treated as a cantilever extending from a central pier or frame. The enormous bending moments in a cantilevered slab with an area of 100,000 square feet or more make either space-frame or thin-plate construction mandatory and often require that the plate be supplemented by other structural devices. The largest folded-plate roof in the United States, for example, embodies three major innovations in modern structural techniques: the Trans World Airlines hangar at Municipal Airport in Kansas City, Missouri, is covered by a prestressed folded-plate roof combined with suspended construction. Built between 1955 and 1956, the structure was the creation of two leading firms of engineers, Amman and Whitney, and Burns and McDonnell. The hangar measures 420 × 816 feet in plan, and the roof cantilevers outward 160 feet on each side of a central anchor block of concrete. The extreme tension in the upper region of the folded plate is taken by steel cables extending in parallel lines from steel posts in the anchor walls to the outer edges of the roof. Before completion of the structures, these cables were drawn to maximum tension in order to prestress the plate, thereby reducing the quantity of reinforcing necessary to sustain the tension in the concrete and dividing the total load between the rib action of the folds and the tightly drawn steel cables.

The principles of the curved shell and the folded plate were combined in the dome covering the Assembly Hall at the University of Illinois in Urbana, erected between 1961 and 1962 from the plans of Harrison and Abramovitz (Fig. 107). The roof of the circular auditorium and field house is an intricately folded shell, spherical in over-all form and 400 feet in diameter. The radial component in the thrust of the dome is sustained partly by a prestressed ring girder at the periphery and partly by the forty-eight buttresses that slope inward as cantilevers in supporting the seats. The reinforcing cables in the girder were drawn to full tension after the concrete had set in order to subject the girder to a compressive stress that would counteract the normal tension induced by the radial thrust. The structural simplicity of this huge though architecturally undistinguished enclosure, which has a total seating capacity of sixteen thousand people, provides another eloquent expression of the scientific exactitude of large-scale modern building.

Shells, indeed, are the most scientific of structural forms, and they were created out of the desire to achieve the maximum economy and efficiency in concrete construction. Their revolutionary effect on architectural form has thus arisen from a scientific technology aiming at practical and utilitarian ends. The extent of this revolution, however, has often been exaggerated by architectural historians and critics, who tend to regard shells as the fundamental building art of the mid-century. It is necessary to remember that shells and folded plates are not load-bearing structures: they cannot carry superimposed floors or even floor loads, nor can they sustain the traffic on a bridge. They are at present limited to the roofs and sheltering covers of specialized buildings for which domed or vaulted construction is appropriate. For this reason shells can play only a limited role in the structural and aesthetic forms of the multistory blocks that constitute the bulk of our public architecture.

There is a small group of curved surfaces that have found a place as load-supporting elements in the structural systems of multistory buildings, and to a certain degree their behavior is comparable to that of shells. Thin conical diaphragms of concrete, cantilevered from central supports, may be used to carry roofs and floors. The pioneer structure is the roof of the S. C. Johnson and Company's administration building, built at Racine, Wisconsin, between 1937 and 1939 as one of Frank Lloyd Wright's few commercial designs. The roof deck is a series of circular discs that are carried on shallow ribbed saucers of concrete cast integrally with the hollow columns located at their centers. These diaphragms are rigidly interconnected by little beams placed at their quarter points, the whole series thus constituting a continuous system of rigid frames. Since there is no bending at the foot of the column, the member tapers downward to a hinged bearing fixed in the floor slab. The diaphragm acts like an inverted dome, with the result that the circumferential area is under tension and is accordingly reinforced with ring bars. Wright and his engineering assistant, Wesley Peters, adopted the same principle on a more extravagant scale for the Johnson company's research tower, constructed beside the earlier building between 1947 and 1950 (Fig. 108). The primary bearing element in the tower is a double-walled hollow cylindrical core, which carries fourteen floors alternately circular and square in outline. The individual floor is the flat upper plate of a hollow diaphragm the lower element of which has the same saucer-like form as the roof supports in the administration building. Much of the reinforcing in the diaphragm is concentrated at the periphery, where there are high tensile stresses arising from the tendency of the member to spread out while bending downward like an umbrella.

Spiral and helical shapes have been employed in the minor functional details of buildings for many centuries, especially for stairways, which can be squeezed into very small compass by wrapping them around posts or circular walls. In modern engineering work spiral segments appear in long bridges on curving alignments and in the scroll casing that carries the turbine water of hydroelectric power plants. Progress in concrete technology eventually made it possible to build spiral or helical stairways and ramps as free-standing elements. The idea of constructing a multistory building in which the floors are merged into a continuous self-supporting spiral is another product of Wright's fertile imagination. He was given the opportunity to try his hand in the new quarters of the Solomon R. Guggenheim Museum in New York, but the conservatism of the city's Department of Housing and Building forced him to modify his original daring plan. He proposed in 1945 to build the exhibition rotunda as a free-standing spiral with six levels at any vertical section. When the museum was constructed between 1956 and 1959, however, he was forced to introduce a number of supporting elements, chief of which are eleven downward-tapering radial slabs that help to carry the ramp load to the footings (Fig. 109). But Wright succeeded in largely hiding these vertical bearing members, so that the upward-expanding spiral appears to revolve freely around the central skylight area.

Hollow cylindrical cores, circular slabs, semicircular cantilevers, and helices are combined in the 600-foot towers of Marina City in Chicago (Fig. 110). These unique structures, designed by the architectural and engineering firm of Bertrand Goldberg Associates, are part of a complex that includes an office building, a bank, a television studio, restaurants, shops, and a marina. Construction of the group was initiated in 1960; all parts were opened for use in 1964, except the studio, which was completed in 1967. Each of the two sixty-story towers is divided into two clearly visible parts, distinguished by function as well as construction. The lower part is a garage in which the parking and access ramp is a helical slab rising continuously from the elevated terrace to a level equivalent to the eighteenth floor. A two-story break separates this part from the forty-two apartment floors that occupy the remainder of the height. The total vertical load of the ramp and the floors and the horizontal wind loads are divided between the two rings of columns, an inner and a peripheral one, and a hollow cylindrical core, which thus functions as a shear wall as well as a primary bearing member. The loads of the helical ramp and the level floors are distributed between the columns and the core by means of radial and circumferential girders. The floor slab is cantilevered outward between columns in a ring of balconies with semicircular outer edges, the repetitive pattern of these scalloped or flower-like shapes greatly intensifying the aesthetic power of the apartment buildings. Other varieties of vault and shell form are embodied in the office building and the studio theater of Marina City. The office enclosure is carried high above the terrace and the single-story bank on a linear series of groin vaults supported in turn by two rows of long slender columns. This compact arrangement of superimposed volumes represents a sound functional solution to the problem of maximum utilization of a small site. The theater is roofed by a steel-ribbed hyperbolic-paraboloidal shell with an unusually deep curvature. The entire Marina City complex thus embraces on an unparalleled scale all the fundamental techniques of modern concrete construction, some of which have been carried to the highest stage of development they have so far reached in American building art.

Chapter Twenty-two | Revivals of Old Materials in New Forms

Our history of building techniques in the present century has focused exclusively on large structures, for which steel and concrete are the only materials capable of sustaining the loads. Since the leading structural innovations have come in this domain, we may lose sight of the fact that the overwhelming majority of small buildings continue to be erected out of the oldest materials by thoroughly traditional means. The detached single-family house, for example, is nearly always constructed with an internal balloon frame of studs, joists, and rafters covered by a wall of masonry veneer or wood sheathing. The footings and foundation walls are concrete poured from a self-propelled mixer, but the rest of the construction process is largely a product of the ancient handicraft skills of the carpenter and the mason. Non-structural materials such as glass, insulating fibers, roof webbing, and the various metals, plastics, and ceramics of the utilities are prepared by mass-production methods, but on the domestic scale they are always installed by hand. Although this division between steel and concrete construction, on the one hand, and timber and masonry, on the other, seemed permanently fixed by the time of World War I, the never ending search for improvements in the economy of building soon led to the revival of old materials in forms suitable for wide-span structures. Most remarkable has been the revival of wood as a primary structural material, in the shape either of small pieces that can be assembled by hand or of very large built-up members.

The return of wood to a position of economic importance in commercial buildings came in the mid-twenties with the importation from Europe of the lamella, or diagrid, system of construction. In this technique a domed or vaulted framework is built up of a great number of small uniform pieces (lamellae) arranged in a dense pattern of parallelograms or triangles, the whole curved into the necessary spherical or cylindrical shape. Structures of this kind are closely related to geodesics since the loads of both must be transmitted chiefly by compression through the axes of the pieces. Their great economic advantage lies in the easy assembly of small, similar, mass-produced components by simple hand-and-tool techniques. Since connections are usually made by bolting, the diagrid frame has inadequate rigidity and must be tied transversely and longitudinally to prevent the members from being pulled apart by horizontal thrusts and by the action of the wind. The common practice of staggering the joints increases the rigidity, although this is severely limited under tension. The light frame represents a particularly effective way of exploiting the elasticity and the balanced tensile and compressive strength of wood.

The first large structure to embody the new principle was unfortunately an ephemeral work that did not survive the single use for which it was designed. The temporary auditorium constructed to house the 1928 convention of the Democratic Party at Houston, Texas, consisted of three parallel halls that were roofed by cylindrical vaults carried on lamella frames with bolted connections (Fig. 111). The three vaults, the central one of which spanned 120 feet, rested on hinged joints that were fixed to the tops of wooden columns located at the edges of the enclosures. Any one vault was thus designed to act like a series of parallel and contiguous two-hinged arches. The extremely low cost of the Houston convention hall was its most attractive feature. The need for stringent economy in the depression, followed by the steel shortage of the war years, greatly stimulated the revival of timber construction, and diagrid structures of wood began to multiply among buildings that did not have to meet the maximum requirements of fire resistance. At the end of World War II, the diagrid was adapted to steel construction, which considerably broadened its range of uses.

The light construction of lamella frames, as in the case of geodesics and shells, limits them to the role of covers carrying only the dead weight of the roof and the lateral forces of wind. If wood is to be reintroduced to sustain the higher loads of modern commercial and industrial buildings and to offer greater resistance to fire, it can be used only in traditional systems of heavy framing in which the members are in various ways protected from rapid combustion. The old frames of sawed timbers were rapidly disappearing around the turn of the century, but a generation later they were to be reborn through the structural development of two closely related inventions. The practice of making plates of wood by gluing thin sheets together and finishing them with a rich veneer was developed by French craftsmen in the late eighteenth century and imported into England by military prisoners during the Napoleonic wars, but it was well into the present century before builders saw the structural possibilities of this laminated material, which we now call plywood. A derivative of plywood is the glued laminated timber, invented by the German engineer Otto Hetzer in 1907. With modern high-strength glues as bonding agents, these built-up forms have at least the strength and physical properties of monoxylic pieces, and when the angle of the grain is varied from layer to layer, the strength may be increased beyond that of the homogeneous wood.

The ready availability of wood following the development of harvesting practices by the lumber industry, along with the low cost of bending, cutting, and shaping operations, eventually brought about the wide adoption of laminated timber for structural purposes, and it was again the depression of the thirties that provided the major stimulus. The first large enclosure with a roof frame of laminated timbers is the Municipal Auditorium at Jamestown, North Dakota, constructed in 1937 from the plans of Edward Tufte and Gilbert Horton. The vaulted roof of the auditorium, spanning 120 feet, is supported by a series of timber arches in the form of hollow-box ribs of a kind originally invented by J. H. Keefe. The laminae at the top and bottom of the rib extend across the full width of the piece, whereas those at the sides are short enough to leave a central opening along the axis of the member.

Extensive testing of laminated members in the laboratories of universities and lumber companies and the wartime shortage of metal led to such a vigorous exploitation of timber construction that for certain kinds of building it became a serious competitor of steel and concrete. The development of new wood preservatives and of bonding agents derived from proteins and synthetic resins give the timber members a high enough resistance to fire to allow their use in public buildings. The heavier laminated timbers are generally used for arch ribs and trusses to support roofs of intermediate span, as in churches, garages, bowling alleys, and the gymnasiums and auditoriums of schools. Plywood has been confined to the relatively thin plates of flanged and box girders and to folded-plate roofs. The great flexibility of wood allows it to be pre-bent into a wide variety of curves, a property that frees the designer from the rigid geometry of rectilinear frames and gives him the opportunity to develop new spatial effects at moderate cost. The technique of laminated timber framing is well exemplified in the gymnasium and auditorium of Thomas Jefferson High School in Port Arthur, Texas, constructed between 1962 and 1963 from the plans of J. Earle Neff and Caudill, Rowlett and Scott (Fig. 112). The roofs of these enclosures form a pair of groin vaults square in plan and measuring 150 feet on a side. The tongue-and-groove sheathing of the roof rests on a highly developed system of main and subsidiary ribs: the primary members, set on the diagonals of the square, span 215 feet with a solid depth of 56¼ inches; the secondary members lie between the diagonals in planes parallel to the sides of the square and range in span from 30 to 150 feet. A system of purlins serves the purpose of distributing roof and wind loads to the secondary arches. The whole framework is much like its steel counterpart in arrangement of members, but the total weight and the cost are very much below what they would have been for a metal structure. The one disadvantage of the timber system is the greater depth of the individual member for a given load and span, which may result in a prohibitive sacrifice of overhead space.

There have been revivals in recent building of ancient structural techniques, but only one has shown promise of playing a continuing role in the building arts. In the years 1936–37 the Farm Security Administration built several co-operative farm communities using the most primitive materials in order to reduce costs to an absolute minimum. In the dormitories of the community at Chandler, Arizona, designed by Burton Cairns and Vernon DeMars, the bearing walls were constructed of adobe, providing a re-statement in the forms of contemporary architecture of the building techniques of the Southwest Indians (see pp. 36–37). The adobe walls not only were capable of sustaining the roof and floor loads of these modest two-story structures but also offered the additional advantage of being excellent thermal insulators. The row houses in the community at Mount Olive, Alabama, designed by Thomas Hibben, utilized the oldest bearing-wall technique known: the walls were constructed of rammed earth in wooden forms, a rare modern example of a method of construction whose origins are lost in the predynastic history of the Mesopotamian peoples.

More useful for the requirements of contemporary building and hence more likely to survive is the reinforced brickwood specified by the architects Harry Weese and Associates for the Columbus Junior High School in Columbus, Indiana (1960–61). The floor and roof loads in this fairly large building are transmitted by steel open-web joists to the rows of brick piers that constitute the bearing members of the outer walls. The corner piers and those supporting the ends of the long trusses, both of which are subject to bending, are constructed of brick laid around a mortar core. The core is reinforced with four vertical rods in the corners, and the brickwork with a succession of hoops laid in every fourth horizontal joint. First proposed by French builders in the late seventeenth century, reinforced masonry represents a stage on the way to the development of reinforced concrete. Brick bearing construction enjoyed a lively resurgence in the 1960's, although it was always used with advanced forms of concrete and steel framing for the interior structures of buildings—a modern counterpart of the iron and masonry structures of the nineteenth century. The great mass of masonry necessary for the walls of the first nineteenth-century skyscrapers has been drastically reduced by the technique of bonding the walls tightly to the reinforced concrete floor slabs, so that all act together as a rigid unit. In spite of the fact that apartment buildings of seventeen stories have been successfully constructed in this way, further development of the art has been hampered in most cities by anachronistic building codes. Although masonry bearing walls, whether plain or reinforced, lie far from the mainstream of contemporary structural art, their revival serves to point up an ironic fact, namely, that the very power of modern technology to multiply new forms obscures the superior virtues of many old and even primitive methods of construction.

Important Dates

1611	First description of English timber framing in the American colonies
1615	Brick construction introduced into the American colonies by this date
1622	Fort, Plymouth, Massachusetts; early example of heavy timber framing
1628	Brick kilns in operation, New Amsterdam
1633	Power-driven saw mill, Virginia
1638	Establishment of New Sweden, Delaware Bay, and introduction of the log house into the colonies
1642	San Estevan Church, Acoma, New Mexico; early work of adobe construction
1650	First shingled house, built by the Dutch on Long Island
1672–75	Castillo de San Marcos, St. Augustine, Florida
1676–95	Senate House, Kingston, New York; early example of dressed stone masonry
1681	Old Ship Meeting House, Hingham, Massachusetts; first truss framing in the American colonies
1720–31	San Jose Mission, San Antonio, Texas
1732–34	Ursuline Convent, New Orleans, Louisiana
ca. 1732–48	Independence Hall (State House), Philadelphia, Pennsylvania
1785	Connecticut River bridge, Bellows Falls, Vermont; first framed bridge in the United States
1792	Salisbury-Newbury Bridge, Massachusetts; first truss bridge in the United States
1793	Slater Mill, Pawtucket, Rhode Island; first textile mill in the United States
1801	Jacob's Creek bridge, Uniontown, Pennsylvania; first suspension bridge in the United States
1811	Thomas Pope's *Treatise on Bridge Architecture*
1814	Power-driven circular saw introduced into the United States
1817	Burr arch-and-truss patented
1819	Canvass White's natural hydraulic cement patented
1820	Town lattice truss patented
1820–22	Chestnut Street Theater, Philadelphia, Pennsylvania; first iron structural members in the United States
1829	Carrollton Viaduct, Baltimore, Maryland; first masonry railroad bridge in the United States

1830	Long truss patented
1833	Balloon frame invented, Chicago, Illinois
	Canfield iron bridge patented
1836–39	Dunlap's Creek bridge, Brownsville, Pennsylvania; first iron bridge constructed in the United States
1837–42	First Croton Dam, Croton River, New York
1839–64	Old Court House, St. Louis, Missouri; first iron dome in the United States
1839	Baltimore and Ohio Railroad bridge, Laurel, Maryland; first iron girder bridge in the United States
1840	Howe truss patented
	Frankford bridge, Erie Canal; first iron truss bridge
1844	Pratt truss patented
1846–47	Baltimore and Susquehanna Railroad bridge, Bolton Station, Maryland; first iron girder bridge in the United States
1847	Daniel Badger's Architectural Iron Works established, New York City
	Squire Whipple's *A Work on Bridge Building*
	Whipple truss patented
1848	Warren truss patented
1848–49	James Bogardus's iron foundry built, New York City
1851–52	Philadelphia, Wilmington and Baltimore Railroad Station, Philadelphia, Pennsylvania; first vaulted train shed on timber trusses in the United States
1851–64	United States Capitol; construction of dome and flanking wings
1852	Bollman truss patented
1854	Fink truss patented
1865–66	Cleveland, Ohio, Union Depot; first iron-framed train shed in the United States
1868–74	Eads Bridge, St. Louis, Missouri
1869–71	First Grand Central Terminal, New York City
1869–83	Brooklyn Bridge, New York City
1871	Artificial Portland cement patented in the United States
	Prospect Park bridge, Brooklyn, New York; first concrete bridge in the United States
1871–76	Ward house, Port Chester, New York; first reinforced concrete building in the United States

1876–77	Kentucky River bridge, Cincinnati Southern Railway; first rail cantilever bridge in the United States
1878–79	Missouri River bridge, Chicago and Alton Railroad; first steel truss bridge
1884	Bear Valley Dam, San Bernardino, California; first arch dam in the United States
1884–85	Home Insurance Building, Chicago, Illinois; first iron- and steel-framed curtain-walled skyscraper
1885	Ernest L. Ransome's first reinforced concrete building
1886–88	Hotel Ponce de Leon, St. Augustine, Florida; first large multifloor concrete building in the United States
1887–89	San Mateo Dam, San Mateo Canyon, California; first concrete dam in the United States
1889	Alvord Lake Bridge, Golden Gate Park, San Francisco, California; first reinforced concrete bridge in the United States
1889–90	Second Rand McNally Building, Chicago, Illinois; first steel-framed building
1889–93	Colorado River dam, Austin, Texas; first masonry hydroelectric dam in the United States
1890–91	Manhattan Building, Chicago, Illinois; first building in the United States with a fully wind-braced iron and steel frame
1894	Melan system of reinforced concrete arch patented
1898	Reinforced concrete rib and girder bridges introduced in the United States
1902–03	Ingalls Building, Cincinnati, Ohio; first reinforced concrete skyscraper
1902	Bureau of Reclamation established
1903	Multiple-buttress dam invented
1903–10	Pennsylvania Station, New York City
1903–13	Second Grand Central Terminal, New York City
1905	Lake Park bridge, Milwaukee, Wisconsin; possibly first concrete-arch bridge reinforced with Kahn trussed bars
1908	Turner flat-slab concrete construction patented
1909	Lafayette Avenue bridge, St. Paul, Minnesota; first flat-slab concrete bridge
	Multiple-arch dam invented
1911–13	Woolworth Building, New York City

1911–15 Tunkahannock Creek Viaduct, Delaware, Lackawanna and Western Railroad

1913 Orpheum Theater, New York City; first rigid-frame construction in steel in the United States

1914–16 Hell Gate Bridge, New York City

1914–17 Ohio River bridge, Chicago, Burlington and Quincy Railroad; longest simple-truss span

Ohio River bridge, Chesapeake and Ohio Railway, Sciotoville, Ohio; longest continuous-truss span

1920 Factory, Electric Welding Company of America, Brooklyn, New York; first welded steel frame

1922 Concrete rigid-frame bridge introduced in the United States, Westchester County, New York

1926–28 Coolidge Dam, Gila River, Arizona; first multiple-dome dam

1927–31 George Washington Bridge, New York City

1928 Convention Hall, Democratic Party, Houston, Texas; first wide-span diagrid frame in the United States

1928–29 Highway bridge, Mount Pleasant, New York; first steel rigid-frame bridge

1928–31 Bayonne Bridge, Bayonne, New Jersey–Port Richmond, Staten Island, New York

1929 Canalization of the Ohio River completed

1929–31 Empire State Building, New York City

1930–31 Westinghouse Memorial Bridge, Pittsburgh, Pennsylvania

1931–36 Hoover Dam, Colorado River, Arizona-Nevada

1933 Tennessee Valley Authority established

Travel and Transport Building, Century of Progress Exposition, Chicago, Illinois; first suspended construction for buildings

1933–37 San Francisco-Oakland Bay Bridge, California

Golden Gate Bridge, San Francisco, California

1933–43 Grand Coulee Dam, Columbia River, Washington

1934 Brook Hill Farm Dairy pavilion, Century of Progress Exposition, Chicago, Illinois; first shell vault in the United States

1934–35 Hayden Planetarium, New York City; first shell dome in the United States

1937 Vierendeel truss bridge introduced in the United States, Los Angeles, California

	Municipal Auditorium, Jamestown, North Dakota; first laminated-timber construction in the United States
1937–39	S. C. Johnson and Company administration building, Racine, Wisconsin; first concrete-diaphragm construction
1938	Water clarifier tank, St. Paul, Minnesota; first prestressed concrete structure in the United States
1946	Aluminum Corporation of America bridge, Massena, New York; first aluminum bridge
1947	Fuller geodesic dome patented
1949–50	Walnut Lane bridge, Philadelphia, Pennsylvania; first prestressed-girder bridge in the United States
1954–57	Mackinac Strait Bridge, Mackinaw City, Michigan
1954–59	St. Lawrence Seaway
1959–61	Kips Bay Plaza Apartments, New York City; University Apartments, Chicago; first load-bearing screen walls of concrete
1959–64	Verrazano-Narrows Bridge, New York City
1960–64	Marina City, Chicago, Illinois
1962–66	Gateway Arch, St. Louis, Missouri
1963–65	Civic Center, Chicago, Illinois
1963–68	Morrow Point Dam, Gunnison River, Colorado; first thin-arch double-curvature dam in the United States
1964–67	Poplar Street bridge, St. Louis, Missouri; first orthotropic-girder span in the United States
1965–70	Hancock Building, Chicago; first steel braced tubular cantilever skyscraper

Suggested Reading

The literature on the history of building techniques as distinct from architectural form is extremely thin, a reflection of the fact that the history of technology has become a serious scholarly discipline only in recent years. The close association between contemporary architectural form and the underlying structure has led to a new interest among critics and historians in the structural basis of building, with the result that the poverty of writing on the subject is gradually being alleviated. There are a number of books on the history of bridges, but many are primarily pictorial, and even the best are far from adequate in their treatment. The history of civil engineering exists only in a few pioneer volumes that are mostly introductory in nature and more like outlines than systematic analytical works.

GENERAL WORKS ON AMERICAN BUILDING

The only books dealing strictly with the evolution of building techniques and structural engineering in the United States are Carl W. Condit, *American Building Art: The Nineteenth Century* and *American Building Art: The Twentieth Century* (New York: Oxford University Press, 1960, 1961). They are marked by certain weaknesses and omissions, but they represent the pioneer attempt to treat the native building arts as part of the history of technology. Some additional material, presented in connection with European development, may be found in the chapters on building, civil engineering, and transportation in Charles Singer *et al.* (eds.), *A History of Technology* (London: Oxford University Press, 1954–58), Vols. IV and V. A useful pioneer work more restricted in scope is Richard S. Kirby and Philip Laurson, *The Early Years of Modern Civil Engineering* (New Haven, Conn.: Yale University Press, 1932). *American Building: The Historical Forces That Shaped It* (2d ed.; Boston: Houghton-Mifflin Co., 1966), by James M. Fitch, has more to say on techniques than most histories of architecture, but the emphasis is social and economic rather than technical.

Among general histories of bridge construction the best is David B. Steinman and Sara Ruth Watson, *Bridges and Their Builders* (2d ed.; New York: Dover Press, 1957). Wilbur J. Watson's *Bridge Architecture* (New York: W. Helburn, 1927) is a handsome pictorial work with a brief descriptive text. *Bridges: A Study in Their Art, Science and Evolution* (New York: William E. Rudge, 1929), by Charles S. Whitney, covers much the same ground as the Watson volume but is more analytical and comprehensive in its descriptions. A German classic with extensive material on American practice is Georg C. Mehrtens, *Vorlesungen über Ingenieur-Wissenschaften,* Part III: *Eisenbrückenbau* (Leipzig: W. Engelmann, 1908 [Vol. I], 1920 [Vol. II]), but it is a technical work. The most thorough history of the science of structure and materials is Stephen P. Timoshenko, *A History of the Strength of Materials* (New York: McGraw-Hill Book Co., 1953), the full understanding of which requires a knowledge of the calculus and differential equations.

The Colonial Period

In the extensive literature on colonial American architecture the only book that deals systematically with building techniques is Harold R. Shurtleff, *The Log Cabin Myth* (Cambridge, Mass.: Harvard University Press, 1939). The comprehensive architectural history by Hugh Morrison, *Early American Architecture* (New York: Oxford University Press, 1952), leans heavily on the Shurtleff book for technical material but includes some details not in the earlier work. A valuable body of material for students is the collection of drawings made by the Historical American Buildings Survey and preserved in the Library of Congress. Two short monographs on special aspects of building construction have recently appeared: Harley J. McKee, "St. Michael's Church, Charleston, 1752–62: Some Notes on Materials and Construction," *Journal of the Society of Architectural Historians,* XXIII (March, 1964), 39–43; and Samuel Wilson, Jr., "Louisiana Drawings of Alexandre de Batz," *ibid.,* XXII (May, 1963), 75–89.

Some material on Philadelphia building exists in various historical works. An account of the Carpenters' Company may be found in Charles E. Peterson, "Carpenters' Hall," in Luther P. Eisenhart (ed.), *Historic Philadelphia* (Philadelphia: American Philosophical Society, 1953). Works concerned with early Philadelphia building are the following: J. Thomas Scharf and Thompson Westcott, *History of Philadelphia, 1609–1884* (Philadelphia, 1884); George B. Tatum, *Penn's Great Town* (Philadelphia: University of Pennsylvania Press, 1961), a superbly illustrated and documented work on the entire history of Philadelphia building but, like most architectural histories, short on structural details. The construction of plank houses is covered in Walter R. Nelson, "Puzzle of the Plank Houses," *New Hampshire Profiles,* XV (March, 1966), 34–40, 52–53, 56. The leading authority on the techniques of colonial building is Lee H. Nelson, architect of the National Park Service at Philadelphia, but unfortunately his voluminous reports and drawings on the structural system of Independence Hall ("Historic Structures Reports on Independence Hall" [1961, 1962]) remain in mimeographed and photostat form. Various papers presented at the National Park Service Historic Structures Training Conferences (Philadelphia, 1961, 1962, 1963) are useful, chiefly the following: Lee H. Nelson, "Eighteenth Century Framing Devices" (1961) and "Nail Chronology as an Aid to Dating Old Buildings" (1962); and by various authors, "Early American Brick Masonry and Restoration of Exterior Brick Walls" (1963). A comprehensive guide to the European works available to colonial builders is Helen Park, "A List of Architectural Books Available in America before the Revolution," *Journal of the Society of Architectural Historians,* XX (October, 1961), 115–30.

From The Revolution to the Civil War

There are no books dealing specifically with the construction of timber-framed and masonry buildings in this period, and the scattered monographs are confined to special aspects of the subject. The comprehensive work on the architecture of much pre-Civil War masonry building is Talbot Hamlin, *Greek Revival Architecture in America* (New York: Oxford University Press, 1944), but the structural characteristics are dealt with sporadically and only in a descriptive way. "A Re-examination into the Invention of the Balloon Frame," by Walker Field, Jr., in *Journal of the Society of Architectural Historians,* II (October, 1942), 3–29, represents the most thorough inquiry into this important invention. Agnes Gilchrist's *William Strickland, Architect and Engineer* (Philadelphia: University of Pennsylvania Press, 1950), supplements Hamlin's book in the area of Greek Revival building. Stewart L. Grow's *A Tabernacle in the Desert* (Salt Lake City: Deseret Book Co., 1958) is the only book to describe the genesis and construction of the Mormon Tabernacle. Carl W. Condit's "The Mormon Tabernacle," in *Progressive Architecture,* XLVII (November, 1966), 158–61, provides an analysis of the structural system. Philadelphia building, which continues to be important throughout much of the nineteenth century, is treated in the book by Tatum, previously cited on p. 296, and in Theo B. White (ed.), *Philadelphia Architecture in the Nineteenth Century* (Philadelphia: University of Pennsylvania Press, 1953).

Certain biographical histories are useful for light on the early practice of engineering: Daniel H. Calhoun, *The American Civil Engineer, 1792–1843* (Cambridge, Mass.: M.I.T. Press, 1960); David Stevenson, *Sketch of the Civil Engineering of North America* (London: J. Weale, 1838); Charles B. Stuart, *Civil and Military Engineers of North America* (New York: D. Van Nostrand Co., 1871).

Timber truss bridges have been treated in several books and long articles of uneven quality. Richard S. Allen's *Covered Bridges of the Northeast* and *Covered Bridges of the Middle Atlantic States* (Brattleboro, Vt.: Stephen Greene Press, 1957, 1959) tend to be anecdotal, local-color histories, but they also include a considerable body of useful facts and many illustrations. Llewellyn N. Edwards' *A Record of the History and Evolution of Early American Bridges* (Orono, Me.: University of Maine Press, 1959) is a more genuinely historical treatise, though oddly disorganized. "A History of the Development of Wooden Bridges," by Robert Fletcher and J. P. Snow, in *Transactions of the American Society of Civil Engineers,* XCIX (1934), 314–408, is a good analytical work by trained engineers. *The Early Years of Modern Civil Engineering* (New Haven, Conn.: Yale University Press, 1932), by Richard S. Kirby and Philip G. Laurson, puts the subject in its general historical setting. Lee H. Nelson's *A Century of Oregon's Covered Bridges, 1851–1952* (Portland: Oregon Historical Society, 1960) is a good monograph for illustrations and basic data. The pioneer classics on the theory of the timber truss are Hermann Haupt, *The General Theory of Bridge Construction* (New York: D. Appleton Co., 1851), and Squire Whipple, *A Work on Bridge Building* (New York: D. Van Nostrand Co., 1847).

Iron framing for buildings is beginning to receive careful attention from historians. The chief sources are Daniel D. Badger, *Illustrations of Iron Architecture, Made by the Architectural Iron Works of the City of New York* (New York: Architectural Iron Works, 1865), and James Bogardus, *Cast Iron Buildings: Their Construction and Advantages* (New York, 1856). Turpin C. Bannister, "Bogardus Revisited," *Journal of the Society of Architectural Historians,* XV (December, 1956), 12–22, and XVI (March, 1957), 11–19, and "The First Iron-Framed Buildings," *Architectural Review,* CVII (April, 1950), 231–46, are admirable monographs. Sigfried Giedion's *Space, Time and Architecture* (3d ed., Cambridge, Mass.: Harvard University Press, 1954), the classic on the history of modern architecture, contains some material on early iron construction. Esmond Shaw's *Peter Cooper and Wrought Iron Beams* (New York: Cooper Union, 1960) is a valuable monograph on the origin of wrought-iron beams in American building. A. W. Skempton's "Evolution of the Steel Frame Building," *Guilds Engineer,* X (1959), 37–51, deals with European and American development.

"Early Metal Space Frame Investigated," by Paul D. Frohman, in *Progressive Architecture,* XLI (December, 1960), 164–71, is a thorough analysis of the iron frame of the United States Capitol dome.

The development of iron truss bridges and their timber prototypes is examined in several early works, which may serve as a basis for the comprehensive treatment the subject deserves. F. B. Brock's "Truss Bridges," *Engineering News,* IX (1882), 371 ff., and X (1883), 5 ff., is the general title of a series of articles on nineteenth-century truss-bridge patents. Theodore Cooper, in *American Railroad Bridges* (New York: Engineering News Publishing Co., 1889), traces the development from the first timber trusses by means of an analytical text and excellent reprints of patent drawings. Karl Culmann's "Der Bau der eisernen Brücken in England und Amerika," *Allgemeine Bauzeitung,* XVI (1851), 69–129, and XVII (1852), 163–222, is a classic that deserves to be translated and published as a book. Early iron bridges are considered in Kirby and Laurson, *op. cit.* (previously cited on p. 298). *Skeleton Structures, Especially in Their Application to the Building of Steel and Iron Bridges* (New York: D. Van Nostrand Co., 1867), by Olaus M. F. E. Henrici, is a treatise that dates from the pioneer phase of the art. Robert M. Vogel's "The Engineering Contributions of Wendel Bollman," *United States National Museum Bulletin* CCLX (1964), 77–104, is a detailed monograph on the work of the Baltimore and Ohio's famous bridge engineer, and his "Roebling's Delaware and Hudson Canal Aqueducts," *Smithsonian Studies in History and Technology* X (1971), is an equally thorough treatment of Roebling's little-known pioneer essays in suspension bridge construction.

From 1865 to 1900

The early development of the iron- and steel-framed skyscraper in Chicago has received careful study, but the New York phase still remains to be investigated. Useful technical works on both New York and Chicago are William H. Birkmire, *Architectural Iron and Steel, and Its Application in the Construction of Buildings* and *Skeleton Construction in Buildings* (New York: John Wiley & Sons, 1892, 1894); and Joseph K. Freitag, *Architectural Engineering, with Special Reference to High Building Construction* (2d ed., New York: John Wiley & Sons, 1901). Carl W. Condit's *The Chicago School of Architecture* (Chicago: University of Chicago Press, 1964) and Frank A. Randall's *History of the Development of Building Construction in Chicago* (Urbana, Ill.: University of Illinois Press, 1949) between them cover the Chicago achievement in detail. "The Structural System of Adler and Sullivan's Garrick Theater Building," by Carl W. Condit, in *Technology and Culture,* V (Fall, 1964), 523–40, is a detailed structural analysis of a skyscraper classic. J. Carson Webster's "The Skyscraper: Logical and Historical Considerations," *Journal of the*

Society of Architectural Historians, XVIII (December, 1959), 126–39, and Winston Weisman's "New York and the Problem of the First Skyscraper," *ibid.,* XII (March, 1953), 13–21, emphasize respectively the structural-functional and the economic aspects of the early skyscrapers. The specific subject of the origin of caisson foundations for buildings is treated in Donald L. Hoffman, "Pioneer Caisson Building . . . ," *ibid.,* XXV (March, 1966), 68–71.

The only work on the history of railroad stations, Carroll L. V. Meeks, *The Railroad Station, An Architectural History* (New Haven, Conn.: Yale University Press, 1956), pays some attention in passing to the construction of train sheds. *Buildings and Structures of American Railroads* (New York: John Wiley & Sons, 1904), by Walter L. Berg, is a useful source on late nineteenth-century practice. For the great age of the rail terminal, the most authoritative work is John Droege, *Passenger Terminals and Trains* (New York: McGraw-Hill, 1916).

Early long-span bridges are considered in the works previously cited by Brock, Cooper, Kirby, and Mehrtens (pp. 299, 298, 296 respectively). Henry G. Tyrrell's *The History of Bridge Engineering* (Chicago: privately published, 1911) carries the subject a little farther but is not always reliable on historical data. The suspension bridge has commanded more attention than the trussed and arched forms, although Ellet's work still remains to be fully investigated. Hamilton Schuyler's *The Roeblings: A Century of Engineers, Bridge-Builders, and Industrialists* (Princeton, N.J.: Princeton University Press, 1931) and David B. Steinman's *The Builders of the Bridge* (New York: Harcourt, Brace & Co., 1945) cover the work of the Roeblings, the latter being more concerned with their bridges. A detailed monograph on a single important bridge (Eads) is Calvin Woodward, *A History of the St. Louis Bridge* (St. Louis: G. I. Jones & Co., 1881).

Literature on the early history of concrete construction scarcely exists. The only general history, Robert W. Lesley, *History of the Portland Cement Industry* (Chicago: International Trade Press, 1924), is out of date and more concerned with material and manufacture than with structures. *Reinforced Concrete Buildings* (New York: McGraw-Hill Book Co., 1912), by Ernest L. Ransome and Alexis Saurbrey, is confined largely to Ransome's own work. "The Ward House: A Pioneer Structure of Reinforced Concrete," by Ellen W. Kramer and Aly A. Raafat, in *Journal of the Society of Architectural Historians,* XX (March, 1961), 34–37, is a short article that omits many of the essential characteristics of this important structural achievement. Josef Melan's *Plain and Reinforced Concrete Arches* (New York: John Wiley & Sons, 1917) deals mainly with Melan bridges. *The Design and Construction of Dams* (New York: John Wiley & Sons, 1899), by Edward Wegmann, is a textbook that describes early concrete structures.

SINCE 1900

The extensive literature on modern architecture includes a number of books treating in varying degree the structural basis of twentieth-century building. Among these are the works previously cited by Fitch, Giedion, and Randall (pp. 295, 298, 299 above). An encyclopedic work on modern building design is Talbot Hamlin (ed.), *The Forms and Functions of Twentieth Century Architecture* (4 vols.; New York: Columbia University Press, 1952), but it is uneven and in places superficial. Books concerned with high buildings are the following: Alfred C. Bossom, *Building to the Skies* (New York: The Studio, 1934); Francisco Mujica, *The History of the Skyscraper* (New York: Archaeology and Architecture Press, 1930); Earl Shultz and Walter Simmons, *Offices in the Sky* (Indianapolis: Bobbs-Merrill Co., 1959); William A. Starrett, *Skyscrapers and the Men Who Build Them* (New York: Charles Scribner's Sons, 1928). These tend to be popular and descriptive: Mujica is best on the development of construction, and Shultz and Simmons emphasize economic and managerial aspects. Leonard Michaels, in *Contemporary Structure in Architecture* (New York: Reinhold Publishing Corp., 1950), deals with European as well as American work but is not concerned with historical evolution. Monographs that provide good illustrations of structural details are Ludwig Hilberseimer, *Mies van der Rohe* (Chicago: Paul Theobald, 1956), and George Nelson, *The Industrial Architecture of Albert Kahn, Inc.* (New York: Architectural Book Publishing Co., 1939). The wind bracing of skyscrapers is treated historically in Carl W. Condit, "The Wind Bracing of Buildings," *Scientific American* 230 (February 1974), 92–105.

A number of recent books provide a non-technical treatment of modern structural theory and practice. The best of these are the following: Mario Salvadori and Robert Heller, *Structure in Architecture* (Englewood Cliffs, N.J.: Prentice-Hall, Inc., 1963); Curt Siegel, *Structure and Form* (New York: Reinhold Publishing Corp., 1962); Eduardo Torroja, *The Philosophy of Structures* (Berkeley: University of California Press, 1958); William Zuk, *Concepts of Structure* (New York: Reinhold Publishing Corp., 1963). Torroja's volume is the only one with historical material.

For railroad terminals, Meeks (already cited on p. 300) carries the subject up to the mid-century decade. The New York terminals are treated in William Couper, *History of the Engineering, Construction, and Equip-*

ment of the Pennsylvania Railroad New York Terminal and Approaches (New York: Isaac H. Blanchard Co., 1912); Edward Hungerford, *Men and Iron: The History of the New York Central* (New York: Thomas Y. Crowell Co., 1938); and William D. Middleton, *Grand Central: The World's Greatest Railway Terminal* (San Marino, Cal.: Golden West Books, 1977).

In the literature of bridges, there are a few non-technical works that deal primarily with twentieth-century achievements. The American Institute of Steel Construction's *Prize Bridges, 1928–1956* (New York: A. I. S. C., 1958) is mainly a pictorial record with data on type, span length, and cost. Georg C. Mehrtens (previously cited on p. 296) emphasizes twentieth-century work in his Vol. III (1923). Elizabeth B. Mock, in *The Architecture of Bridges* (New York: Museum of Modern Art, 1949), is concerned with visual form rather than structural features. *Miracle Bridge at Mackinac* (Grand Rapids, Mich.: Eerdmans Co., 1957), by David B. Steinman and John T. Nevil, is a popular account of how a big suspension bridge is built. Joseph B. Strauss's *The Golden Gate Bridge* (San Francisco: Golden Gate Bridge and Highway District, 1938) is a sober chief engineer's report to the board of directors. *A Decade of Bridges* (Cleveland: J. H. Jansen, 1937), by Wilbur J. Watson, is a supplement to his earlier *Bridge Architecture* but more thorough in treatment. Mock and Watson deal with concrete as well as steel bridges.

Modern concrete building has received a good deal of attention from architectural historians, but bridges continue to be neglected. The following examine, in passing, structure as well as visual form: Peter Collins, *Concrete: The Vision of a New Architecture* (New York: Horizon Press, 1959); Jürgen Joedicke, *Shell Architecture* (New York: Reinhold Publishing Corp., 1963); Francis K. Onderdonck, *The Ferro-Concrete Style: Reinforced Concrete in Modern Architecture* (New York: Architectural Book Publishing Co., 1928); Aly A. Raafat, *Reinforced Concrete in Architecture* (New York: Reinhold Publishing Corp., 1958). Except for Onderdonck, these books emphasize European work more than American. A good review of the current state of the art is "Concrete Technology in the U. S. A.," *Progressive Architecture,* XLI (October, 1960), 158–85; a survey of progress in the century is "A Review of Fifty Years of Progress and Experience," *Concrete and Constructional Engineering,* LI (January, 1956), entire issue.

There is a considerable literature on dams but no general historical work. *Register of Dams in the United States* (New York: McGraw-Hill

Book Co., 1958), edited by T. W. Mermel, provides a comprehensive list with essential data on function, type, date, and size. The work of the Bureau of Reclamation is covered in United States Department of the Interior, Bureau of Reclamation, *Dams and Control Works* (Washington, D.C.: Government Printing Office, 1954), a thorough and well-illustrated treatise that ought to be brought up to date. Navigation dams of the Corps of Engineers are described in United States Department of the Army, Office of the Chief of Engineers, *The Mississippi River between the Missouri River and Minneapolis, Minnesota,* and *Report upon the Improvement of the Ohio River* (both published in Washington, D.C.: U. S. Army, Corps of Engineers, 1958). The literature on TVA is voluminous. The concept of regional planning and the functioning and organization of the authority are described in Gordon R. Clapp, *The TVA: An Approach to the Development of a Region* (Chicago: University of Chicago Press, 1955), and David Lilenthal, *TVA: Democracy on the March* (New York: Harper Bros., 1944). The Tennessee Valley Authority's *Annual Report* (Knoxville: TVA, 1953), the twentieth anniversary report, includes a good history of the background and genesis of the authority. John H. Kyle's *The Building of TVA: An Illustrated History* (Baton Rouge: Louisiana State University Press, 1958) is the best non-technical description of the physical plant. More thorough and technical descriptions are contained in Tennessee Valley Authority, *Engineering Data: TVA Water Control Projects,* Technical Monograph No. 55, Vol. 1 (Knoxville: TVA 1954), and *Engineering Data: TVA Steam Plants,* Technical Monograph No. 55, Vol. 2 (Knoxville: TVA, 1955). Construction of steam generating facilities has been so rapid in recent years that only the *Annual Report* can keep the reader abreast of the expanding physical plant.

The chief periodicals that provide source material on the structural character of individual works are *Civil Engineering, Engineering News, Engineering News-Record, Engineering Record, Railroad Gazette, Railway Age, Sanitary Engineer,* and *Transactions of the American Society of Civil Engineers,* but these are professional engineering journals designed for technically trained readers. *Railroad Gazette* and *Sanitary Engineer* are no longer published, and the *News* and the *Record* were merged to form the present *Engineering News-Record.*

Credits for Illustrations

1. From L. F. Salzman, *Building in England down to 1540* (London: Oxford University Press, 1952).
2. From Harold R. Shurtleff, *The Log Cabin Myth* (Cambridge, Mass.: Harvard University Press, 1939).
3. From Hugh Morrison, *Early American Architecture* (New York: Oxford University Press, 1952).
4. Drawing by National Park Service.
5. Drawing by National Park Service.
6. From *Journal of the Society of Architectural Historians.*
7. From *Transactions of the American Society of Civil Engineers.*
8. From Hugh Morrison, *Early American Architecture* (New York: Oxford University Press, 1952).
9. From drawings of the Historical American Buildings Survey.
10. From Charles Singer *et al.* (eds.), *A History of Technology* (London: Oxford University Press, 1954–58).
11. Drawing by Illinois Central Railroad.
12. Courtesy Central Vermont Railway.
13. Drawing by Melvin Brown and Le Grand Haslam; courtesy Church Building Committee, Church of the Latter-Day Saints.
14. From *Transactions of the American Society of Civil Engineers.*
15. Courtesy McGraw-Hill Book Company.
16. From *Engineering News.*
17. From Henry G. Tyrrell, *History of Bridge Engineering* (Chicago: Privately printed, 1911).
18. Courtesy McGraw-Hill Book Company.
19. From Talbot Hamlin, *Greek Revival Architecture in America* (New York: Oxford University Press, 1944).
20. Courtesy Historical Society of Pennsylvania.
21. From drawings of the Historical American Buildings Survey.
22. Courtesy McGraw-Hill Book Company.
23. Photo by Erie Railroad.
24. Courtesy Museum of Modern Art, New York.
25. Photo by Wayne Andrews.
26. Photo by Charles R. Payne.
27. From drawings of the Historical American Buildings Survey.
28. Drawing from the Fine Arts Commission, Washington, D. C.
29. Photo from Public Buildings Service.
30. From *Engineering News.*
31. From *Engineering News.*
32. Drawing by author.
33. Photo from Cheney Collection, United States National Museum.
34. From *Transactions of the American Society of Civil Engineers.*
35. Photo by Union Pacific Railroad.
36. Courtesy McGraw-Hill Book Company.
37. Photocopy by Keller Studio; courtesy Cincinnati Public Library.
38. From *Architectural Forum.*
39. From William H. Birkmire, *Skeleton Construction in Buildings* (New York: John Wiley & Sons, 1893).

40. From *Inland Architect and News Record.*
41. Photocopy courtesy William Rudd, Historical American Buildings Survey.
42. From Frank A. Randall, *History of the Development of Building Construction in Chicago* (Urbana: University of Illinois Press, 1949).
43. Photo by Kenneth Oberg.
44. From *Industrial Chicago* (Chicago: Goodspeed Publishing Co., 1891).
45. From *Industrial Chicago.*
46. Photo by Richard Nickel.
47. Courtesy New-York Historical Society.
48. Photo by Reading Company.
49. Photo by Delaware, Lackawanna and Western Railroad.
50. Model and photo by Smithsonian Institution.
51. Drawing by author.
52. Courtesy St. Louis-San Francisco Railway.
53. Courtesy McGraw-Hill Book Company.
54. Courtesy St. Louis Chamber of Commerce.
55. From Henry G. Tyrrell, *History of Bridge Engineering* (Chicago: Privately printed, 1911).
56. Courtesy New York Department of Public Works.
57. Courtesy Portland Cement Association.
58. From *Transactions of the American Society of Civil Engineers.*
59. From *Engineering News.*
60. Courtesy Portland Cement Association.
61. From Ernest L. Ransome and Alexis Saurbrey, *Reinforced Concrete Building* (New York: McGraw-Hill Book Company, 1912).
62. Courtesy Recreation and Park Department, San Francisco, California.
63. From *Engineering Record.*
64. Courtesy Museum of the City of New York.
65. Courtesy New York Central Railroad.
66. Courtesy New York Central Railroad.
67. From *Engineering News.*
68. From *Engineering News-Record.*
69. Photo by Richard Nickel.
70. Photo by *Chicago Tribune.*
71. Photo by Ezra Stoller; courtesy Skidmore, Owings and Merrill.
72. Photo by Frank Lotz Miller; courtesy Curtis and Davis.
73. Photo by Donald Eldridge; courtesy First National Bank of New York.
74. Photo by Hedrich-Blessing.
75. Photo by Chicago, Burlington and Quincy Railroad.
76. Courtesy Bethlehem Steel Company.
77. Photo by Harvey Patteson; courtesy Southern Pacific Railroad.
78. Photo by New Jersey State Department of Highways.
79. Drawing by author.
80. Courtesy Los Angeles Department of Public Works.
81. Courtesy New York, New Haven and Hartford Railroad.
82. Photo by Port of New York Authority.
83. Courtesy Museum of the City of New York.
84. Photo by Port of New York Authority.
85. Photo by California Department of Public Works.

86. Photo by Roy Flamm; courtesy San Francisco Chamber of Commerce.
87. Photo by *New York Times.*
88. Courtesy General Motors Corporation.
89. *Top:* from *Engineering News; bottom:* from Max Bill, *Robert Maillart* (Zürich: Verlag für Architektur A. G., 1949).
90. Photo by Richard Nickel.
91. Courtesy Delaware, Lackawanna and Western Railroad.
92. Courtesy Allegheny County Department of Works.
93. Photo by California Department of Public Works, Division of Highways.
94. Courtesy City Planning Commission of Cincinnati.
95. Photo by Philadelphia Department of Public Works.
96. Photo by United States Army.
97. From Bureau of Reclamation, *Dams and Control Works* (Washington, D.C.: Government Printing Office, 1938).
98. Photo by Bureau of Reclamation.
99. Photo by Bureau of Reclamation.
100. Photo by Bureau of Reclamation.
101. Photo by Bureau of Reclamation.
102. Photo by Jerome Berman.
103. Photo by Tennessee Valley Authority.
104. Photo by Tennessee Valley Authority.
105. Courtesy Portland Cement Association.
106. Courtesy Portland Cement Association.
107. Courtesy Douglas Fir Plywood Association.
108. From Leonard Michaels, *Contemporary Structure in Architecture* (New York: Reinhold Publishing Corp., 1950).
109. Photo by Ezra Stoller; courtesy Solomon R. Guggenheim Museum.
110. Photo by Richard Nickel.
111. From *Engineering News Record.*
112. Photo by Judge Pictures; courtesy Koppers, Inc.

Index

[All bridges, buildings (including grain elevators, houses, monuments, reservoirs, stadiums, and windmills), canals, dams, expressways and parkways, railroad stations, and railroads are indexed by type or name under those general entries.]